CONSERVATIVE CONSERVATIONIST

CONSERVATIVE
conservationist

Russell E. Train and the Emergence
of American Environmentalism

J. BROOKS FLIPPEN

LOUISIANA STATE UNIVERSITY PRESS
BATON ROUGE

Published by Louisiana State University Press
Copyright © 2006 by Louisiana State University Press
All rights reserved
Manufactured in the United States of America
First Printing

Designer: Michelle A. Garrod
Typeface: Whitman, Gotham
Printer and binder: Edwards Brothers, Inc.

Library of Congress Cataloging-in-Publication Data

Flippen, J. Brooks,–
   Conservative conservationist : Russell E. Train and the emergence of American
environmentalism / J. Brooks Flippen.
       p. cm.
   Includes bibliographical references and index.
   ISBN-13: 978-0-8071-3203-6 (cloth : alk. paper)
   ISBN-10: 0-8071-3203-9 (cloth : alk. paper)
1. Train, Russell E., 1920– 2. Cabinet officers—United States—Biography. 3. Judges—United
States—Biography. 4. United States. Dept. of the Interior—Officials and employees—Biography.
5. Conservationists—United States—Biography. 6. Environmentalists—United States—Biography.
7. Environmentalism—United States—History—20th century. 8. Environmental policy—United
States—History—20th century. 9. Conservation of natural resources—United States—History—
20th century. 10. World Wildlife Fund—Biography. I. Title.
E840.8.T73F55 2006
333.72092—dc22

                                                                                    2006010524

Published with support from the Louisiana Sea Grant College Program, a part of the National Sea
Grant College Program maintained by the National Oceanic and Atmospheric Administration of the
U.S. Department of Commerce.

For Maya and Emily

CONTENTS

Illustrations follow page 86.

ACKNOWLEDGMENTS

"The harder I work," Samuel Goldwyn once quipped, "the luckier I get." In researching and writing this book, I have learned the truth in this simple wisdom. Good fortune has followed all of my efforts. At every step of the process I have received outstanding assistance and I owe a great deal of thanks as a result. Russell and Aileen Train were kind enough to house and feed me at their home in Florida, Russell patiently sitting for hours of taped interviews. Without access to his personal papers, no amount of hard work could have produced this book. The good people at the World Wildlife Fund endured an annoying researcher who tended to monopolize their copying machine. My dear friend Steve Jones and his family kindly took me in when I visited Washington. Southeastern Oklahoma State University provided a sabbatical from my teaching, Vice President for Academic Affairs Jesse Snowden undoubtedly proving a strong advocate on my behalf. I wish to thank the many environmentalists and government officials who agreed to interviews and Dr. Leonard Bruno in the Manuscript Division of the Library of Congress. My student assistant, Christina Gray, always performed professionally the many tedious and mundane tasks I assigned her. If hard work produces good luck, her future is bright. A Washington-based researcher, Athena Angelos, helped with the photographs while the archivists at the National Archives never once let me down. Guiding my hard work was an outstanding editor, Alisa Plant, who had the unenviable task of answering my many e-mails. This book would not have been possible without her and the other great professionals at Louisiana State University Press. Last but hardly least, my wife, Celeste, suffered through draft after draft, always with a keen eye for my many mistakes. Why, I wonder, does she put up with me? In the end, one might question my own hard work. No one, however, should doubt my good luck in dealing with such fine friends and colleagues.

CONSERVATIVE CONSERVATIONIST

INTRODUCTION
Medal of Freedom

Several months past his seventy-first birthday in June 1991, Russell Errol Train received a phone call from the White House. The news was a pleasant surprise. President George Bush planned to award him the Medal of Freedom, the nation's highest civilian honor. It was the culmination of a career that spanned half a century and established Train as one of the nation's most powerful environmentalists. "From the standpoint of conservation leaders the world over," one environmental advocate applauded, "Russell Train exemplifies the closest thing we have to royalty."[1]

Although obviously gratifying, the news should not have surprised Train. The recipient of thirteen honorary degrees, he already owned more than a dozen prestigious awards from environmental organizations, professional associations, and academic institutes.[2] Twice the National Wildlife Federation awarded him its Conservationist of the Year Award, one of only six individuals so recognized. Organizations such as the Natural Resources Council of America, the National Academy of Sciences, the American Society of Civil Engineers, and the United Nations concurred, at various times lauding him for his contributions. Elected a fellow of the American Academy of Arts and Sciences, Train long served as a trustee or director of numerous nonprofit organizations, several of which he helped establish.[3]

"A list of Russ Train's government offices," the *Washington Post* gushed, "would be thick enough to stop a raging bull."[4] After service in World War II, Train worked as a congressional staffer and Treasury Department tax attorney before gaining an appointment as one of the nation's youngest federal judges. While still on the bench, he established the African Wildlife Leadership Foundation (AWLF), dedicated to preparing newly independent Africans for conservation. He soon resigned his judgeship and became president of the Conservation Foundation. As his reputation grew, he played a vital role in the formation of an American affiliate of the World Wildlife Fund (WWF). In time, President Lyndon Johnson appointed him to the National Water Commission, which led to membership in several critical federal task forces.

Recognizing the young Republican as a political asset in a world in which environmentalism promised a new constituency, President Richard Nixon returned Train to government service full-time with an appointment as Under Secretary of Interior. A year later when the National Environmental Policy Act passed, Nixon offered Train a new challenge, naming him the first chairman of the Council on Environmental Quality (CEQ). In this capacity Train helped guide federal policy during one of the most important periods in American environmental history.[5] Benefiting from Nixon's interest in foreign affairs and the spread of environmentalism around the globe, Train led the American delegation to the famous 1972 U.N. Conference on the Human Environment in Stockholm, Sweden. He played a critical role in early American-Soviet environmental alliances, a key component of Nixon's policy of detente, and helped establish the World Heritage Trust. As the environmental movement and American policy matured, Train left CEQ after three years to become the second administrator of the Environmental Protection Agency (EPA), a position he held for four additional years, throughout the tumultuous "energy crisis" and the Presidency of Gerald Ford.

When Jimmy Carter won election in 1976, Train's government service ended and he returned to the private sector, ending his career as president and chairman of WWF. He continued to accept public service appointments on an ad hoc basis, including environmental mediation and service on the United Nations' Brundtland Commission. WWF prospered during his tenure despite a growing backlash to environmentalism that threatened many of his accomplishments. "Since Russ Train took over the helm of the WWF," friend and former attorney general Elliot Richardson remarked, "it has risen faster than everything but Microsoft."[6] Under Train's direction, WWF instituted a unique "Debt-for-Nature Program" and helped establish several independent conservation trusts. It issued a series of key reports and created a National Commission on the Environment. No doubt in recognition of Train's stature in the world environmental community, biologists working in the wet forests of Costa Rica named a newly discovered species of ant *Pheidole traini*. The other honors aside, here was a distinction that no one else in government or the environmental community could claim. "Your achievements are laudable," Bush's Secretary of Interior Lynn Martin wrote to her Republican colleague. "Your dedication knows no equal."[7]

If Train questioned whether his resume warranted the Medal of Freedom, he might have recognized that his friendship with George Bush aided his cause. Train's father, an admiral, was a friend to Bush's father, Senator Prescott Bush. When the younger Bush was a Republican congressman from Texas, he and his wife Barbara regularly played mixed-doubles tennis with Train and his wife Aileen. Train helped in several of his friend's campaigns and, in 1975, his family joined the Bushes in the People's Republic of China, a country only beginning to feel the thaw of the Cold War and where Bush served as one of America's first liaisons. Staying at the opulent Peking Hotel, the Trains ate almost every meal with the Bushes. "There is no tourism in the normal sense," Train recalled in his diary, "but Bush has the privilege of having personal guests."[8] Such connections allowed them to visit the Forbidden City, the Great Wall, the Summer Palace, the Ming tombs, and several local communes. Four of Bush's children toured with them, including the eldest, twenty-nine-year-old George W. Bush, in one of his first overseas trips. The Bush children played basketball with the locals in one small step for detente. "You and Barbara were terrific to take us on and to devote so much time to us despite having your own kids there," Train thanked Bush afterwards, solidifying a friendship that both assumed would last a lifetime.[9] In fact, only months before offering Train the Medal of Freedom, Bush invited the noted conservationist for drinks at the White House. Operation Desert Storm had just begun and American troops were liberating Kuwait from Iraqi occupation. As the group discussed Baghdad and encouraged the commander-in-chief, Bush fielded phone calls from National Security Advisor Brent Scowcroft. The Iraqis were firing scud missiles at Israel, the concerned but calm President explained. Train sat quietly listening, fate making him a witness to history.[10]

Word spread quickly of Train's Medal of Freedom, and friends and colleagues rushed to offer congratulations. "So often I read about awards and it's not clear to me why a particular recipient has been chosen," wrote one acquaintance. "Clearly this award is meritorious."[11] Congratulations came from every corner of the world, a reflection of both Train's international reputation and impact. Letters from South Africa, Nepal, Ecuador, Pakistan, and other far-flung locales arrived, several in languages that Train could not understand.[12] The letters were from corporate executives, academicians, scientists, attorneys, and dozens of environmentalists of all stripes, each recalling his or her own story of how Train warranted the recognition.

On the day of the award ceremony, November 18, 1991, Train arrived early at the White House to clear security. With him came Aileen, his four children, and his two brothers, spouses in tow. One grandson also attended, representing the newest generation of the family. The ceremony took place in the late morning in the East Room with a luncheon following in the State Dining Room. Joining Train in receiving the medal were nine other prominent individuals, including former First Lady Betty Ford, writer and pundit William F. Buckley, Jr., former Speaker of the House Thomas "Tip" O'Neill, and baseball great Ted Williams. The recipients sat on the first row before a low stage, their family and friends seated behind them. President and Barbara Bush stood at a podium, a marine officer nearby calling each of the ten to the stage and reading the official citations. President Bush followed with a few comments before Barbara Bush placed the medal over each recipient's neck. Train's citation was succinct and noted that he had "helped shape society's growing environmental awareness into sound policy." America, the citation concluded, "honors an ardent conservationist whose efforts help preserve Nature's treasures in this country and around the world."[13] His name called, Train approached the First Lady only to have her joke loudly enough for all in the audience to hear, "Aren't you going to kiss your old friend?"[14] The ceremony might have warranted official protocol but President and Mrs. Bush were not about to let formality keep them from enjoying the success of a friend.

Returning to his seat, Train found the elderly Williams dozing. "You're up," Train quipped as he nudged the famous hitter, clearly enjoying the moment no less than the Bushes. At the lunch that followed the banter was equally collegial. Train sat on Mrs. Bush's left with Tip O'Neill on her right. "It's too bad [Iraqi leader] Saddam Hussein decided George would never attack and not stand by what he said he would do," the First Lady joked. "If he had asked me, I would have told him." When one of her large costume jewelry earrings dropped into his crotch, Train recalled, "its retrieval gave rise to a number of ribald comments on sexual harassment." Aileen and the rest of the family sat interspersed among the famous dignitaries and politicians. Sitting next to Train's daughters, for example, were Scowcroft and Secretary of State James Baker. The conversations flowed easily, more festive than formal. "After all," Train recalled, "we all knew each other."[15]

One might have forgiven Train, who was well into the normal retirement age, if he had taken a few days off to enjoy his laurels. He had other plans, however, for within hours of the ceremony he was on board a plane bound for London. He had a dinner appointment at Saint James Palace with fellow WWF trustee Prince Bernhard of the Netherlands. Sitting next to Bernhard's wife, Princess Juliana, Train enjoyed conversation no less collegial than at his White

House luncheon. "[Juliana] really enjoys her drinks," Train recalled. "At one point she remarked that she was almost drunk, but she wasn't." The next day it was lunch at Buckingham Palace with Prince Philip, the Duke of Edinburgh. Train, it appeared, enjoyed a boundless energy and impressive connections, both so influential to the success of his career and cause. In the end, the laurels he received should have surprised no one.[16]

Train's Medal of Freedom reflected his long career but hardly told the whole story of his life or the significance of his impact. His decades of historic service, both successes and failures, did not take place in a vacuum; public sentiments and concerns were dynamic, their shifts influencing both Train and those he encountered. The politics of it all was just as fluid. Changing political alliances and priorities always defined for Train the realm of the possible. In a sense Train's many years reflected the evolution of American environmentalism and the politics that surrounded it. His life story is a tale of modern American environmental history.

Train lived through it all. In the days of his youth, the assumptions of the Progressive Era still dominated conservation. "Elite groups," in the words of historian Hal Rothman, drove the movement. The poor often suffered the consequences of environmental degradation but, faced with the daily struggle to survive, had little time for anything beyond putting food on the table. The wealthy, by contrast, not only had the education to appreciate the breadth of the problem but the time and means for reform. Able to retreat to areas of natural beauty, they developed an appreciation alien to those lower on the socioeconomic ladder. Born into a family of privilege, Train exemplified the traditional conservationist. He had frequent and early exposure to nature, an excellent education, and a love of travel. His interest grew from a vacation safari in Africa, hardly an avocation for the average American. Once convinced of the threat to African wildlife, he worked diligently but relied upon many of the advantages he enjoyed. Progressive conservationists, Rothman also notes, focused on "regional or place-specific issues . . . at the so-called pristine areas instead of communities where people lived."[17] The AWLF grew from the African Safari Club, a group of prominent Washingtonians who shared a love of hunting in East Africa. It did not focus on urban blight in nearby communities but, rather, a romantic, distant, and untamed land.

The gospel of efficiency guided progressive reformers of all stripes, with conservationists no exception. Success demanded enlightened and educated leadership, a reliance upon trained experts employing the best of contemporary scientific management.[18] Train and his colleagues at the AWLF were a case in point. Their goal was to prepare Africans in the postcolonial world to manage wildlife professionally. Like much of progressivism, a certain paternalism was evident, perhaps magnified in Africa given traditional race relations. Nevertheless, such reforms helped address the problems of a rapidly industrializing and urbanizing world, with the AWLF an obvious success. Both the Africans and their wildlife ultimately benefited, the underlying elitist assumptions notwithstanding.

Conservationists were not preservationists. They sought to protect nature for, and not from, man; they did not seek to remove a resource from use but rather to manage it properly to ensure perpetual use. Train and his allies enjoyed hunting and experiencing nature; their goal was not to restrict those that followed but to ensure that the future provided the same opportunities. Train had little appreciation for John Muir and other notable preservationists of his parents' generation, and, surprisingly given his later career, had no interest in such famous preservationist battles as Echo Park Dam, which was taking place even as the AWLF formed. He had no relation to, or opinion of, organizations such as the Sierra Club. His view, like many of his fellow Americans, was not as broad.

Train's view of nature evolved, however, with the majority of the public soon following. By the late 1960s, the costs of unrestrained urbanization and industrialization spawned a new environmentalism concerned with not just traditional conservation but also environmental quality.[19] It recognized the world as a whole and man's impact on it. Learning the importance of habitat to the ultimate survival of the animals he loved, Train soon reflected this growing ecological perspective. One could not divorce population growth, pollution, pesticides, and the myriad of other threats from wildlife. Conservation of African wildlife was important but it was only one symptom of a more complex problem involving the fragility and interdependence of all life. The AWLF began to expand its efforts and Train moved on to the Conservation Foundation, one of the earliest organizations to appreciate this web of life. There he helped spread the alarm, educating the public to the downside of its affluent lifestyle. He did not share the attire—or the youth—of many of the new environmentalists but, like them, questioned traditional assumptions. He may not have recognized it at the time but he stood at the fore of a revolution in American attitudes. Indeed, when Nixon summoned Train to the helm of CEQ, the nation prepared to celebrate its first Earth Day.

If Train's life is a story of environmentalism, it is also a tale of Washington and the political class and culture that guides it. As reflected in the Medal of

Freedom ceremony and his relationship with Bush, Train was always a creature of the nation's capital, operating easily among its power elite. His family's prominence afforded him important connections, perhaps the most influential currency in the market of politics, and the chance to learn early the interpersonal and diplomatic skills necessary to operate in such a world. He was a "master politician," according to Vice President Nelson Rockefeller. "You have had about the toughest job in Washington but you have stayed in there," he told Train.[20] Like most in Washington, Train was ambitious, never one to miss an opportunity to advance his career. He did so, however, without alienating potential competitors or opponents. In the words of Nixon's Secretary of Labor James D. Hodgson, "One is not long on the Washington scene before one comes to expect lucidity and grace from the words of Russ Train."[21] According to former National Park Service director George Hartzog, Train "was so nice that even the opposition didn't manifest itself with any degree of openness or intensity."[22] Train routinely sent colleagues letters of congratulation or condolence. He frequently kept meetings informal and hosted dinners after them. Substantive discussions took place over glasses of wine as well as conference tables. In the words of one experienced Washingtonian, "Washington parties are serious affairs."[23] Nowhere was this more true than in regard to Train.

The Washington Train valued was a town of bipartisanship. Republicans and Democrats, conservatives and liberals alike, worked together more often than not. Legislators from both sides of the aisle joined in the Washington social scene, which lent itself to compromise. The result was a policy of moderation. It was the era of Dwight D. Eisenhower, the champion of consensus politics and pragmatic centrism. The ideological conservative Robert Taft had just died; the press had exposed the partisan demagogue Joseph McCarthy as a disgrace; and the Democrats had abandoned much of the class-based populist rhetoric of the New Deal era.[24] "When the chips were down in those days," presidential advisor David Gergen recalled, "Republicans and Democrats stood firm together."[25] Born into a family of Republicans, Train always remained loyal to the Grand Old Party. Like Eisenhower, however, he believed that government could still play a positive role and remained open to new ideas. Years later one of Train's colleagues, John Whitaker, described him as "an environmentalist first, Republican second."[26] In Train's Washington, there was no problem in being both.

Any political consensus, of course, did not hold. The turbulent 1960s, Vietnam, and Nixon eclipsed the relative moderation and calm of the Eisenhower era. Overt partisanship once again became the rule of the day, making Train's dual role more problematic. Recognizing Nixon as a partisan animal, Train appealed to his political instincts to continue the environmental gains that both Republicans and Democrats had worked so hard to accomplish. Such a tactic was somewhat ironic for a man who valued compromise and working with the opposition; his actions arguably augmented the very partisan tensions he denounced. Nevertheless, savvy to the winds of politics, he knew what now resonated. Nixon unleashed an unprecedented environmental agenda to win what he saw as a new constituency, an agenda that Train heavily influenced. Train thus helped engineer revolutionary changes in the way America dealt with her natural bounty.

This did not satisfy Nixon. The electoral boon that he anticipated never materialized. The Democrats moved to solidify their advantage on the issue, always promising more than the conservative in the President could tolerate. The environmental vote was wide but not deep, Nixon now assumed, the other momentous events of the day always overshadowing the issue. Environmental voters, forever favoring his opponents, did not appreciate his administration's actions. Worse, he risked alienating his natural conservative base. Industry increasingly protested the new restrictions, which Nixon now concluded hampered the economy. The solution was clear; his administration should "get off the environmental kick."[27]

The backlash had begun, complicating Train's efforts. No longer was Train, now frequently on the defensive, always fighting to advance the environmental agenda. Increasingly frustrated and ambitious as ever, he moved to EPA. The task, however, did not grow any easier; the emerging "energy crisis" and the economic malaise of the 1970s sapped the strength of the environmental movement and, in a sense, Train's clout in the White House. In the broader context, the liberalism of the Great Society era had begun to spark the inevitable backlash. A new, rejuvenated conservatism grew, which flourished in the years that followed. Among many Republicans, pragmatism gave way to a more ideologically driven distrust of government. Extremes seemed to dominate. Train's moderate stance, which recognized new limitations, was occasionally timid, frustrating environmentalists who demanded a stronger defense against the shifting tides. "Train knew what he had to do to remain in good standing with the Nixon administration," former Sierra Club President Laurence Moss noted.[28] Former Interior Secretary Stewart Udall, a onetime admirer, announced that he was "disappointed" in Train's performance. Added another environmentalist, "We sometimes wish [Train] would make stronger public statements, but he won't."[29] Rancor grew throughout government. "Apparently you can't be friends anymore with somebody who is of a different party," Arizona Senator John McCain soon complained. "There was no place for a moderate," former diplomat Christian Herter concurred.[30] Train remained in the middle, and the Republican Party he knew and the Washington he loved were on the wane.

Given this trend, perhaps Train's return to the private sector was inevitable. The new Republican ascendancy culminated in what historian Samuel Hays terms the "Reagan antienvironmental revolution." President Ronald Reagan, Hays argues, "rejected environmentalists as legitimate participants in the give-and-take of public affairs." He assumed environmentalism a phenomenon easily "swept aside by vigorous presidential leadership."[31] In the words of historian Jeffrey Stine, the new administration "sobered the environmental movement, confronting them with a new political landscape hostile toward their core values."[32] Reagan was undeniably popular in declaring government "part of the problem, not part of the solution," but his antiregulatory assault posed a significant threat to the nation's environment. Nevertheless, the environmental movement did not vanish even if it was obstructed. During the new era of conservative rebirth, private environmental organizations flourished, perhaps as a backlash of their own. For Train's part, the WWF enjoyed considerable success. Now joined with the Conservation Foundation, it became part of a network of affiliated organizations with offices in over seventy countries. With a broad and diverse funding base including donations from individuals, corporations, and foundations, it today enjoys net assets of over $146 million. Although it occasionally worked with dictatorial regimes, the WWF, fully cognizant of the perilous ecological balance in play, still has spent $83 million over the course of its existence to protect wildlife and wild lands.[33] Its efforts help combat the global threats of uncontrolled deforestation, overfishing, toxic chemicals, and climate change—and the hand of Russell Train has been there from the beginning.

As with many of his environmental colleagues, now more lobbyists than bureaucrats, Train's work on behalf of WWF did not divorce him completely from the life of government service he had known. His commitment to his cause, if not his love of Washington, kept him surrounded by politics. This was a task more difficult for Train than for many of his fellow environmentalists because he shared the party affiliation of the deregulators now controlling so much of the public agenda. Torn between party loyalty and an irritation at its new direction, Train was weak in denouncing Reagan. One article in the *Washington Post* aside, his relative silence given the magnitude of the environmental onslaught was unfortunate, even as his work in the private sector was impressive. In the words of former Sierra Club President Michael McCloskey, Train had "a moral obligation" to criticize his party's leadership more forcibly. "He betrayed his cause, for as a Republican his comments would have carried more weight."[34] Train acknowledged that "we will never achieve excellence in government by trashing it at every opportunity," but many environmentalists wished that he had stressed the argument more vigorously.[35]

Train remained the loyal Republican, voting for Reagan and, in time, his friend George Bush. Much to his dismay, however, even Bush increasingly embraced the neoconservative revolution within his party, a development that Train described as both "mystifying and frustrating." Train privately voiced displeasure, employing both his connections and diplomatic skills but, in public, remained largely supportive even as much of the administration's record suggested otherwise. Finally, with the election of Bush's son, George W. Bush, whom he had met a quarter of a century earlier in China, Train could take it no more. Once again he spoke out, this time more forcefully and in several venues, criticizing the new President's environmental plans and forever straining his friendship with the elder Bush, who valued loyalty more than anything. "Sadly, while we continue to run into the Bushes from time to time," Train wrote, "the easier friendship of earlier days seems gone."[36] In 2004, well into his ninth decade of a life loyal to the Republican Party, Train did something he had never done before. He voted for John Kerry, the Democratic candidate. Indeed, in the end, Train was an environmentalist first and Republican second.

Now in the winter of his life, Train appreciates his Medal of Freedom and other accolades even as he harbors apprehension over the future of his cause. He reflects upon a half-century of public service that has taken him to every corner of the globe and provided an entrée to the world's elite. It was not just George Bush and Prince Bernhard who called Train a friend. At the vortex of modern American politics, living in one of the most important cities in the world, he was a witness to history. From World War II to the global war on terrorism, from the 1960s counterculture to Reaganomics, Train was in or near the halls of power. The struggle to protect America's natural environment continues, not always with the success for which Train and others have so long labored. Nevertheless, environmental history will undoubtedly record the importance of Russell Errol Train in that endeavor. His legacy is secure.

1

A Family of Prominence, a Career of Promise

Thomas Jefferson predicted in 1813 the rise of a "natural aristocracy" in the young American republic, families whose "virtue and talents" afforded them a level of prominence.[1] The Trains of Massachusetts, who prospered even as Jefferson lived, proved a case in point. The vanguard of the family, twenty-two-year-old John Trayne, arrived in the Massachusetts Bay Colony in 1635, one of the first English Americans. Braving the hazards of a months-long trans-Atlantic journey with his bride and thirty-two other hardy souls, John quickly purchased land, a house, and a sawmill, evidence that like many in the "Great Migration" of Puritans during that era, he did not arrive destitute. The cost of the journey alone was more than the annual salary of a typical English yeoman.[2] This did not mean that life was easy; the five-year-old colony numbered only several thousand and the vast wilderness was at its doorstep. Despite the hazards of the New England winters, the hostility of Indians and the ever-present threat of disease, John's industriousness served him well. He prospered and bequeathed to his family not only specie but "houses, barns, orchards, and meadowlands."[3]

The generations of the Train family that followed helped turn the pages of American history. They fought the French in the French and Indian War. Several were minutemen ready at a moment's notice to defend hearth and home, while others witnessed the "shot heard around the world" at Concord Bridge. A member of the Train family suffered with George Washington at Valley Forge while another helped defeat the British at the Battle of Saratoga, the crucial victory that convinced the French to join the Revolution. After independence Russell Train's great-great-grandfather, Charles Train, broke from family tradition and embraced the Baptist faith. In doing so he joined the Great Awakening then still sweeping America, a movement more common among less-educated frontiersmen than Harvard graduates like him. His faith instilled in him a strong belief in both public education and the separation of church and state, principles upon which the fledgling democracy depended. His second wife introduced the name Russell, derived from an English ancestor, to the family.[4]

Charles's New England existence was relatively modest compared to those of other Train descendants, who included one of Boston's leading shipowners and merchants. One, George Francis Train, an eccentric world traveler who met, among others, President Zachary Taylor, British Prime Minister Lord Henry Temple Palmerston, and French Emperor Louis Napoleon, claimed to be the inspiration for Jules Verne's 1873 novel *Around the World in Eighty Days*. His love of travel was hardly unique among Trains, as future generations proved. Several members of the family went into the ministry, an occupation that did not provide an exorbitant salary but still brought its own prestige. Not all descendants, however, shared the industriousness common to the clan. Given to drink, one "remittance man" survived on family gratuities and inheritance.[5]

Ambition characterized the family. Early members served in the Massachusetts legislature and Russell Train's great-grandfather, Charles Russell Train I, was a congressman at the outbreak of the Civil War. He had been a delegate to the first Republican National Convention in 1856, proving that the Train family's Republican roots extended back to the party's birth. When war broke out, the politician became a soldier, requesting a commission and agreeing to pay his own expenses. He fought in the Battle of Antietam, known as the bloodiest single day of the conflict, and received a commendation from his commander.[6]

For Russell Train's grandfather, Charles Jackson Train, public service meant a career in the United States Navy. While military service did little to enhance the family's financial situation, certainly not to the level of many of his descendants and relatives, his distinguished career added a new level of prominence to the family. After graduation from the Naval Academy at the conclusion of the Civil War, the new officer toured the world. A newly imperialist America sought to extend her power, allowing Charles to call in ports from St. Petersburg to Istanbul in a transient life that perhaps only his eccentric relative George Francis Train could appreciate. In the years that followed he taught at his alma mater, commanded a cruiser as part of the blockade of Cuba during the Spanish-American War, and even helped plan naval strategy for that brief 1898 struggle. Promoted to admiral, he assumed command of the entire Asiatic fleet in 1905; one of his first duties was to deal with the conclusion of that year's Russo-Japanese War.[7]

His son, Charles Russell Train II, Russell Train's father, continued the tradition of naval service, graduating from the Naval Academy at the dawn of the new century. Six years later, commanding a gunboat on the Yangtze River, he was with his father when the admiral tragically died of a mysterious illness.[8] The young officer's tour of duty continued, involving diplomacy with Chinese viceroys and, in one instance, action to save American missionaries. He harbored no qualms about the righteousness of America's Open Door Policy; like many Americans, he overlooked the subjugation of the Chinese. Rising in rank like his father, his exposure to the famous and powerful was just as extensive. Only months before his father's death, he joined him in escorting Theodore Roosevelt's daughter Alice and her fiancé House Speaker Nicholas Longworth on a cruise to meet the Dowager Empress Tz'u Hsi. In the years that followed, a friendship developed between Alice's family and the Trains.

Not surprisingly, both Trains were strong supporters of President Roosevelt. Indeed, reflecting the policies of Captain Alfred T. Mahan, who argued that naval power determined national power, Roosevelt made a concerted effort to expand and modernize the nation's navy. Steam-powered and armored "dreadnaughts," turrets atop, became the symbol of national pride, naval personnel the new heroes in an emerging age of imperialism.[9] When Roosevelt launched his famous Great White Fleet, an armada of warships ironically painted white to symbolize peace, to circumnavigate the globe, Charles Russell received a command as part of the fleet. As the flotilla departed from Hampton Roads, Virginia, a crowd of thousands on hand, both he and his admiral father greeted President and Mrs. Roosevelt, the latter suggesting to the younger Train that he write her of the fleet's progress. In later years Charles Russell even grew to resemble T.R., sporting a bushy mustache popular at the time and spectacles, often perched near the end of his nose as he enjoyed his games of poker.[10]

Support for Roosevelt only strengthened the Train family's Republican credentials. The family held few Democrats, although as military officers both Trains avoided overt political partisanship, which was considered a violation of military code. Charles Russell was not a complicated man. A former boxer in college, he was a man of action who did not read to any great extent. Fiercely loyal, he could be quick-tempered. Restrained from political activism, he had strong conservative convictions that often emerged at the dinner table.[11] Now a high-ranking officer in his own right, Charles Russell admired the internationalism of Woodrow Wilson but not his progressive domestic agenda. Indeed, during the early years of World War I, Charles Russell served as a naval aide at the American embassy in Rome, there at the request of Thomas Nelson Page, a friend of Wilson's and a member of a prominent Virginia family. In this capacity he met Austrian-Hungarian Emperor Franz Josef. Once Italy joined the war, he assumed command of all the American forces in the country, attending several important military conferences and meeting the Prince of Wales, later Edward VIII. When President Wilson came to Rome following the war's conclusion, thousands of Italians angry over the Treaty of Versailles greeted him shouting, "Crucify Wilson." "They crucified another good man 2,000 years ago," Charles Russell wrote in his journal. Clearly the young officer firmly embraced the Grand Old Party yet was not an ideological reactionary.[12]

The political demise of Wilson in 1920 and the rise of the Republican majority dominant throughout the new decade brought Charles Russell even closer to the halls of power. After receiving the Navy Cross for his war service, he held several high-ranking positions in the Navy Department's Bureau of Navigation, administrative posts broken by a two-year stint in command of a naval vessel. He preferred the sea command but his job in Washington had its own perks. Supervising the naval athletes in the 1924 Paris Olympics, he met Johnny Weismuller, later famous for his portrayal of Tarzan but then a world-record holder in the 400-meter freestyle. He witnessed one of the great moments in Washington sports history—the Senators' surprising victory over the New York Giants in the 1924 baseball World Series. As the years passed he made many friends, including the Italian and Argentine ambassadors to the United States and Secretary of State Frank Kellogg, one of the authors of the Kellogg-Briand Pact of 1928. With air travel exploding in the 1920s, aviators became new national heroes and Charles Russell joined his fellow Americans in admiration. Soon he met such luminaries as Charles Lindbergh, Robert Byrd, World War I ace Eddie Rickenbacker, and Amelia Earhart. An especially close friendship developed with Admiral William Halsey, later famous in World War II.[13]

It was, however, his relationship with President Herbert Hoover that brought Charles Russell closest to power. In 1928 the Navy Department ordered him, then commander of the battleship *Utah,* to escort Hoover and his wife on the final stage of their postelection goodwill tour of South America. Picking up his distinguished guests in Uruguay and transporting them on a three-week journey back to the United States, Charles Russell befriended the President-elect. Hoover was not the remote cold politician so often portrayed in the media, he was surprised to learn, but a warm affable individual given to conversation. Hoover must have thought the same of his host, for just over a year later Hoover requested that Charles Russell serve as his White House naval aide.[14] Serving as a liaison between the President and the Navy Department was not always an easy task, with the latter concerned about Hoover's conservative budget and his proposed limit on naval construction. Nevertheless, Charles Russell handled the job with aplomb, enjoying the opportunity to rub elbows with the rich and famous, if not the strong ceremonial component of the work. He met, among others, Marshall Pétain and Premier Laval of France, the King and Queen of Siam, and Franklin Delano Roosevelt, then New York governor but soon to become Hoover's political adversary. There was no way the disabled governor could handle the stresses of the Presidency, Charles Russell assumed. Upon meeting Calvin Coolidge and his wife, he described the former President as "the usual cold fish" but his wife "a dear."[15] Through it all, Charles Russell's relationship with Hoover grew. For him, detailed discussions of naval policy were all that surpassed the informal picnics, retreats to the country, and Potomac River cruises on the presidential yacht that they both enjoyed. It was a heady time.

Power meant Washington, of course, and by this time the Trains had firmly established themselves in the nation's capital. In fact, despite his absences at sea, Charles Jackson Train had first set down roots in Washington a decade before the birth of Charles Russell, the city then hardly impressing him. "The city is full of politicians and their attendants," he wrote, "and I think I never saw a more disreputable set of men as I meet on [Pennsylvania] Avenue."[16] In time the close proximity to national power—and the insular world of the Washington powerful—proved advantageous for his progeny, who obviously did not share his first impression. There have always been two Washingtons, *Washington Post* columnist Meg Greenfield later wrote: the locals who go about their business as permanent citizens as in any town, and the politicos who pass through, staying only as long as their elected tenure or appointments last but in the interim developing their own elite culture. Attending the same functions and operating within the same mores, these Washington insiders live almost in an inbred world oblivious to the larger context around them. They may hail from different parties or areas of the country, but "they think of themselves nonetheless as kind of a shaggily coherent whole."[17] It was this world to which Charles Jackson Train settled his family, embedding them in a culture that helped define future generations. They were in a unique position; their family were members of the political class and yet unlike most of their elected and appointed political colleagues, they called the District of Columbia their permanent home. There they witnessed administrations come and go, political philosophies and politicians rise and fall, all the while maintaining friendships and surviving for the next day. Crossing both worlds, they were the true Washington insiders.

If nothing else, it was Washington where Charles Russell met his wife, Errol Cuthbert Brown, a beauty with long brown hair and fine features. Her family's connections to the nation's capital dated all the way to 1807, when the town was nothing more than a small village in a swamp off the Potomac. She too traveled among the Washington elite, her ancestors prospering more than the Trains. Among her family were the chaplain of the House of Representatives and United States Senate; one of the founders of what later became George Washington University; and a director of one of Washington's leading banks, the latter also an owner of a boardinghouse that became a popular rendezvous for Jacksonian politicians.[18] Her parents built a large home near the Trains on Connecticut Avenue and thus she and her future husband grew up together, only blocks from the White House. They shared more than their locale, however, for both families had by this time shed their Baptist origins and joined the Episcopal faith. One can only surmise the reason for the change in denominations, but in any event their new allegiance left them more in common with their upper-class contemporaries. The Trains attended Saint John's, the old church situated across Lafayette Park from the White House and a religious home to many presidents. For years the Trains maintained a rented pew there, before the church suspended the practice in the 1960s as elitist. Like her future husband, Errol attended Washington's prestigious private schools and, like most women of her class and day, she graduated to a "finishing school" when Charles Russell began his naval career. They married on June 15, 1908, an event that warranted three columns in the *Washington Post*.[19]

A family soon followed. Errol gave birth to a son just over a year later, Cuthbert Russell Train, and another son, Middleton George Charles Train, arrived in April 1913. Life was not always easy, because a naval officer's salary did not match the income that their family's prominence suggested. The young family initially rented a residence and instituted other cost-saving measures. Nonetheless, in time they rented a second summer home in Jamestown, Rhode Island and, in 1923, purchased their own home, a large three-story Italianate "villa" on Q Street. The price was $17,000, almost four times the average price of a new home.[20] Their efforts sustained them in their more affluent circle, with private schools for the children, membership in Washington's tony Metropolitan Club, and frequent tennis for the talented player Errol. Charles Russell concentrated on his career; as with his father, his success demanded frequent absences from the family. Errol remained the loving wife and doting mother, devoted to her growing brood. Calling her own mother almost every day, she ensured her children a close-knit family. It was this life to which a third son, Russell Errol Train, was born on June 4, 1920.[21]

* * *

Life in Georgetown had its advantages. While his brothers were significantly older, soon to attend boarding school and college, and his father frequently traveling, young Russell Train still enjoyed an active life. In addition to his mother, the child had a nurse and his family a live-in cook. His education began with French lessons at the age of five and progressed to well-respected private schools, the Potomac School through the fourth grade and Saint Albans School through his 1937 high school graduation. For decades these schools have remained among the choices of Washington's elite, and it is not surprising that Train began to develop his own notable associations. His closest friend was Orme Wilson, the son of an ambassador, while other friends included the children of Secretary of Treasury Andrew Mellon, Army Air Force hero Billy Mitchell, and President Harry Truman's Secretary of State Dean Acheson, the latter child coming to Train's sixth birthday party only to howl for the return of his mother.[22]

Train later wrote that he was slow to mature, a "late bloomer."[23] Such might have been the result of his skipping the third grade, an early indication of his intelligence but a decision that left him a year younger than his peers. It might also have been the result of his physical stature. Never particularly strong as a child, he suffered a lot of colds and never excelled as an athlete, the clear determinant of status in adolescent boys. Oddly he tried boxing, but the beatings he endured led him to baseball and he became a fan of the Washington Senators. Clearly his father played a role in the formation of his personality. While his father taught him honesty and loved the boys intensely, he could be stern and quick-tempered. Corporal punishment was not unusual. Train obviously drew from his father a driving ambition that characterized much of his life but learned a different way to deal with the competitiveness that surrounded him. Less combative and more introspective, Train developed a more diplomatic and amicable personality than his domineering elder, who wore his heart on his sleeve. With the intelligence to know when and where to speak, Train learned how to cultivate and maintain friends. His father enjoyed life, but arguably Train enjoyed it more. As he would later comment, "Perhaps I am the antithesis of my father." Together with his notable connections and his status as a Washington insider, it was a trait that served him well for the rest of his life.[24]

While his father may have lacked a deep intellectual bent, Train had other role models. He was close to his aunt's husband, for one, the noted judge Augustus Noble Hand. A cousin of a fellow judge, the aptly named Learned Hand, Augustus Hand graduated from Harvard Law School with straight A's and proceeded to become what the *New York Times* described as "a great jurist."[25] A recipient of numerous honorary degrees and one of the few Democrats in the large family, he served on the United States Court of Appeals for the Second Circuit, one of the nation's highest courts. To Train he was just "Uncle Gus," however, a man who held all the complexity and depth that his father lacked, an imposing man with big bushy eyebrows and a habitual scowl but who loved children and regaled them with stories, poetry, and songs.[26]

Augustus Hand did share one trait with his brother-in-law Charles Russell, which both passed on to the young Train—a love of the outdoors. The stories Hand told the children were of his youth, camping in the Adirondack Mountains with Learned as panthers howled nearby. As the 1920s progressed, Train joined the Hands at their red-brick mansion in the Adirondacks. It must have been quite a contrast to the bustling city life with its buses and streetcars. Not only did Train escape the interminable heat and humidity of the Washington summers but thick forests and fields surrounded the house, "a wonderful playground for a small boy," he recalled. When Train was ten, he spent one summer with an aunt on Cape Cod. Unlike more recent times, it was then even more remote than the Adirondacks. An old house stood on a knoll with fields and orchards nearby and pine woods enclosing the whole. No traffic or noise disturbed the serenity. It was here that Train first discovered the joys of fishing, camping, and swimming, the latter mitigated somewhat by wool trunks with straps over the shoulders and a skirtlike cloak around the bottom.[27] Soon sailing added a love of the water, which persisted throughout his life.

For his part, Charles Russell shared with his kin a love of hunting. In fact, Charles Russell's love of the sport provoked an international incident in 1905 when, overseas, he accidentally shot and wounded a Chinese woman, inciting a local mob and necessitating a rescue by American forces. Most hunting expeditions were not quite as momentous but still enjoyable for the young Train, who undoubtedly valued the time alone with his father as much as the sport. Duck hunting on the Potomac River, pheasant hunting in the country, and later, hunting excursions with his brothers furthered Train's exposure to nature. In time it would alter the trajectory of his life.[28]

For the most part, however, Train continued his maturation in the Washington scene. The same year as Train's Cape Code summer, Hoover invited him to the White House for a photo opportunity and to greet another youngster the same age who had saved a group of children from a snowbound bus in the Midwest. Two years later, as his family prepared for Charles Russell's transfer to the Naval War College in Newport, Rhode Island, the Hoovers invited the

Train boys for a sleep-over at the White House, the better to get them out of their parents' hair. As his brothers slept in a large Victorian double bed in the Andrew Jackson bedroom in the northeast corner of the great house, Train slept in a small, adjacent bedroom. The next morning the three enjoyed breakfast with the First Family on the southern portico. With an incredible view of the Washington Monument across the Ellipse, Train recalled not only the friendliness of his powerful host but also the largest glasses of fresh-squeezed California orange juice he had ever seen.[29] It was power through the eyes of a child.

Hoover's power was short-lived, of course, the Great Depression robbing him of a second term and leading to Franklin Delano Roosevelt's New Deal. The Trains hardly welcomed the change, relating little to the newly destitute. The Depression certainly affected them—cuts in the federal budget mandated for Charles Russell a captain's pay even as he now carried the rank of admiral—but not to the extent of many Americans. With the New Deal came thousands of new government bureaucrats pouring into the District of Columbia, the city in the words of one historian "under siege."[30] Miles of red-brick row houses sprang up in what open spaces remained, and the "temporary" war buildings still left from the Great War were replaced by impressive new structures. It all buoyed the local economy. Charles Russell was not in danger of losing his job, unlike the quarter of the American workforce unemployed in 1933, and young Train was not in danger of forfeiting his education or the advantages he enjoyed. While Charles Russell was in Newport, the Trains did not possess the wealth to join the grand summer society of that town. Nevertheless, this did not stop them from calling on Mrs. Cornelius Vanderbilt, the railroad magnate's wife and the aunt of Train's best friend Orme Wilson.[31] Not once did Train have to hold a summer job; his circumstances were far from the desperation that fueled the political revolution then taking place.

The young Train loved to travel and his family had the means. He enjoyed his summers away from Washington as a boy and, in 1933, traveled by car across the country with his mother and brothers, his father having assumed command of a Pacific fleet. This was not a relatively short excursion along the well-maintained interstate highways of today, but an arduous journey of almost two weeks on roads that often remained gravel. With a running board on the side of their car and their possessions tied to the top, they undoubtedly resembled the infamous "Okies" of the Dust Bowl. It was hard but it opened Train's eyes as no experience before. Relegated as the youngest to the backseat, he felt the monotony of the vast expanse of the Great Plains, which included the memory of prairie dogs that stayed with him the rest of his life. He saw the majesty of the towering Rockies and the beauty of the Pacific coast. It must have impressed the boy, augmenting an appreciation of nature if not whetting an appetite for travel.[32]

Soon he was old enough to travel abroad, a sure sign of status in an age of depression. Beginning in 1934 and extending to 1939, at the outset of World War II, he joined his friend Orme and his ambassador father in Europe, one summer in Berlin, two summers in Prague, and a final summer in Brussels. He may not have appreciated it at the time, but it offered him a front-row seat to history. In 1934 he saw the prominent "brownshirts" in Berlin's streets, their swastikas proudly displayed on their armbands. He was there when the Reichstag burned, the ornate home of the German legislature that Hitler secretly set afire in an attempt to provoke a crisis. He joined the angry crowds that night as large black limousines arrived carrying Nazi dignitaries there to inflame the passions of the crowd. Two years later he witnessed the Berlin Olympics, waves of the Nazi salute and cheers of "Sieg Heil!" greeting Hitler as he entered the stadium, Train and his party conspicuously sitting throughout with their hands in their pockets. He applauded the victory of track star Jesse Owens, the African American who undercut Hitler's claim of Aryan superiority. In 1939 he heard the news of the outbreak of hostilities as he sailed from Le Havre in France to Southampton in England. There he had trouble getting into the harbor; his transport's docking was delayed by mines and barrage balloons placed in defense of the attack the British thought sure to come. Scheduled to play in British tennis tournaments, he and his friend Orme kept advancing as their opponents forfeited, called into military service.[33] As they returned to the United States on an American passenger liner, the two learned of the torpedoing of a similar British ship on virtually the same course, disquieting news that kept all on edge.

Train was now a young man and it was time for him to leave home. Charles Russell hoped that his youngest son would follow his footsteps to Annapolis, ultimately becoming a third-generation admiral; his two older boys had been physically disqualified. Train, however, chose to follow his siblings to Princeton University. He brought to college above-average grades and college board scores, as well as awards for his editorship of his high school newspaper. Nevertheless, his strong legacy and family of prominence undoubtedly played a role in his admission.

His start at Princeton was less than auspicious. Still immature, he took to collegiate life with gusto, and cigarettes, beer, and liquor joined tennis as key components of his free time. In later years Train wrote that his social life was not a success, although a list of his many activities undermines such a claim. He tried polo and joined the Terrace Club, one of the upper-crust "eating clubs." On occasion he ventured into New York City with his roommate, a jazz fanatic whose parents owned a home in midtown. There they went to 52nd Street and Greenwich Village to hear such greats as Count Basie and Duke Ellington (whose father had been employed as a butler to Train's uncle). The Duke and the other black musicians played before the packed all-white audiences, Train and his friends elbowing their way in and nursing a drink at the crowded, smoky bar. The value of an Ivy League education, one might argue, is the associations it develops. For Train, it only expanded his connections. In one instance, for example, he joined with John L. Lewis, Jr., the son of the renowned labor leader, to steal the bell tower from Nassau Hall, a perennial student escapade.[34]

On holidays and vacations, Train attended the parties and debutante balls that were common among the upper classes throughout America and that in Washington were often held at the Sulgrave Club. Here families "presented" their daughters to proper society and, in doing so, provided a socialization process for the elite. The young people probably did not care about this, preferring the opportunity it afforded to mingle with the opposite sex. For Train, who had suffered through dance classes as a youngster, it meant another opportunity to visit the White House, still the home of Franklin and Eleanor Roosevelt. FDR's niece was "coming out" and the President acted as a host for a party in the East Room. FDR enjoyed watching the youthful festivities from his wheelchair. The fact that alcohol was not part of the party did not deter Train. Noting that Roosevelt had his own drink, Train recalled that one of his father's colleagues had since risen to the position of White House chief usher. Train boldly asked him for a drink and was told to wait inside the entrance to the Red Room. In a few minutes a waiter arrived with a tray covered with a cloth napkin, hiding the requested scotch and soda. Clearly Train was learning how Washington worked.[35]

Not surprisingly, his social life came with a cost. The hours at the Nassau Tavern—the "Nass" to students—contributed to poor grades as he began his studies and led to a stern warning from the dean. In one instance, Princeton city police arrested him for trespassing before suspending the charges upon the promise of disciplinary action from university authorities. Even at graduation, hung over from the parties the night before, Train slept through the ceremonies—not exactly a statement of gratitude to his parents, who had driven up for the occasion. For the most part Train slowly matured during his four years at Princeton, however, his indiscretions more an aberration than commonplace.

In the end his parents did indeed have reason for pride. Train buckled down and reversed his lackadaisical attitude and graduated *cum laude*. He majored in political science, and while none of the great political theorists he read influenced him to any significant extent, his growing interest in the politics of the day was evident. Only months before Pearl Harbor, his senior thesis, *The United States versus Japan: A Study of Sea Power in the Pacific*, certainly proved timely. According to his advisor, the thesis constituted a "careful, objective appraisal of the relative chances of the United States and Japan in a war in the Pacific."[36]

The future lay before him, perilous though it may have seemed, and Train still had no long-range goals. He contemplated a career in foreign service and, to encourage such, his father sought to utilize his connections. He sent a copy of Train's thesis to friend and Assistant Secretary of State Norman Armour. It was "prescient," Armour replied. In any event, Train's immediate future was certain. Perhaps as penance for rejecting his father's hopes that he attend Annapolis or national pride in anticipation of a war, he had joined the Army Reserve Officer Training Corps as a freshman, since there was no Navy ROTC then at Princeton. He now owed military service.

First Lieutenant Russell Train reported for duty at the Field Artillery Replacement Training Center at Fort Bragg, North Carolina, in July 1941. There his service proved rather mundane, providing inductees basic training before their assignment at field artillery units around the country. The routine did not last long, however. At home one weekend, he got a call from a friend telling him to listen to the radio: the Japanese had attacked Pearl Harbor. Together with his parents and dressed in uniform, Train walked to the Japanese embassy, where a crowd of several dozen had formed. The crowd, curious but not angry, watched in silence as Japanese diplomats scurried in and out of the door, smoke rising from the chimneys in what surely was the destruction of classified documents. When a newspaper reporter arrived, someone thought a show of disdain proper. "Show your fist," he cried, although few complied.[37]

With the war came transfers—first to Fort Sill, Oklahoma, for a focus on gunnery, then to Fort Hood, Texas, for training in tanks, and finally to Louisiana for maneuvers. Train was a good officer, promoted to major, but his father's prominence undoubtedly helped his career. When Charles Russell visited his son at Fort Hood, the base's commanding officer entertained the admiral royally. Forced into retirement because of disability, Charles Russell lamented the lost opportunity for combat but, ever the good soldier, kept his complaints to himself. The Army knew an opportunity when it saw one, and its public relations department took the occasion of Charles Russell's visit to note that Train, both a grandson and son of an admiral, had in fact joined the Army.[38]

By the time the Army transferred Train to Okinawa, the war was almost over. Die-hard Japanese remained lodged in caves, however, sniping at Americans when they could. Charged with removing the threat, Train's men threw in smoke grenades in an attempt to root them out. If this were not enough, a vicious typhoon hit. Warned of a possible tsunami, an enormous and destructive tidal wave, Train ordered his troops to higher ground. He knew that he would be at the fore of an invasion of Japan, because he had attended several planning sessions. Fortunately, however, the atomic bomb saved him and thousands of others such a bloody campaign. After Hiroshima and Nagasaki, it fell to the likes of Train to receive the surrender of various Japanese units. In one such instance, working with a translator, Train gave a short speech to a surrendering Japanese colonel and his men, and he ceremoniously received the colonel's pistol.[39]

Once his service was over, it was back to civilian life and, for Train, law school at Columbia University. Rejected by Harvard Law School with a comment about his "high friends at court," Train thrived back in New York City. His career goals remained vague, but Train recognized law school as an entrée into a number of professions. With his military experience, he was now mature enough not to let his social life distract him. Reflecting his interest in politics, he decided to join the Legislative Drafting Research Fund (LDRF), a nonprofit entity within the law school dedicated to the premise that writing effective legislation required professional training and skill. Over the years the LDRF had assisted the New York Legislature, among others, and had helped spark the Office of Legislative Counsel in both the United States Senate and House of Representatives. Students considered an invitation to join the LDRF an honor just below the status of law review. Train, with solid but not outstanding grades, was a marginal case. He had befriended Jack Kernochan, however, the senior student member of the LDRF and a man with some pull. It was not the last time Train's connections and interpersonal skills would work to his advantage.[40]

About to embark on his career, Train was well prepared for success. His background had formed his interests, personality, and affiliations that guided him in the years to come. He entertained no thought of a career in natural resources and yet a latent interest in nature and travel existed. Even with his legal studies, he took time to camp at a friend's 5,000-acre property in the Ontario wilderness. After law school graduation, he celebrated with a fishing trip to Acapulco, Mexico, marveling at the porpoises nearby, the tuna leaping out of the water, the birds in hot pursuit, and, of course, the capture of an eighty-pound dolphin.[41] He enjoyed the Washington life and operated easily within its insider culture. He had the innate intelligence, superb education, and important connections that served him well.

It was hard to deny his Republican credentials. Perhaps not the most politically active in college, he still had worked in a campaign to defeat the Democratic mayor of Jersey City, New Jersey. He had voted against the election of Harry Truman and, in law school, had joined the New York Young Republicans. He was, however, too embedded in the Washington culture to be a rabid partisan. He knew and appreciated too many people from both sides of the aisle and recognized the need for bipartisanship. He approved of Truman's use of the bomb and, like his father, worried about the more conservative orientation of Republican Robert Taft, a family acquaintance.[42] More liberal in many of his views of government and religion than others in his party, he was not an ideologue. The 1944 G.I. Bill of Rights, which provided educational aid for veterans, had, in fact, helped him through law school. In one lay sermon several years later at Saint John's Church, he rejected religious dogma. "Christ himself never preached doctrine but taught a way of life," he explained. The country needed "a new tolerance and a new compassion and, perhaps, a new wisdom."[43] His formal education now complete, Train's individual needs were more specific. He needed a new direction.

One could certainly understand if Russell Train, aged twenty-eight and the proud new owner of a Legum Baccalaureus degree, had sought a career in one of the large national law firms. He had lacked for nothing but neither had he possessed the great wealth common among his contemporaries or those equally prominent. He had lucrative offers, interviewing for one job with former Secretary of State Dean Acheson at the prestigious firm Covington and Burling. For Train, however, Washington's political culture in which he felt so comfortable beckoned.[44]

Just before graduation, he put his interpersonal skills and connections to work. He contacted a friend who, in turn, introduced him to Gordon Grand, clerk of the House Ways and Means Committee. One of the oldest committees in Congress, the Ways and Means Committee had primary responsibility for all revenue issues, including aspects of tariffs and trade, the Social Security system, and welfare programs. It ranked, therefore, as one of the most influential committees in Congress.[45] Impressed, Grand introduced him to Colin Stam, the longtime Chief of Staff to the Joint Committee on Internal Revenue Taxation. This committee, established by the Revenue Act of 1926, included senior members of the Ways and Means Committee as well as the Senate Finance Committee. It provided professional staff to its two constituent committees, neither with similar support. Fat, bald, and unkempt, Stam was an odd character but, like Grand, he was impressed by Train, most notably because of Train's work on the LDRF. He offered Train a job upon graduation with surprisingly good pay. A rather lonely bachelor, Stam was an unusual fellow to befriend the popular Train, but his decision was a crucial first step in launching a career that had ramifications far beyond taxation policy.[46]

The work as a staff member was tedious. Initially sharing a small office in the Old Senate Office Building with two economists, Train proofread tax legislation, corresponded with the staff of the Treasury Department, researched issues of note, and, perhaps most importantly, helped to hammer out differences between Senate and House taxation bills. In one instance his assignment was to write a legislative history of the oil depletion allowance, a percentage of the gross income from production intended as compensation for the resource's depletion. Train's report, however, did little to justify tax breaks for the industry, which committee members from oil-producing states expected. Not surprisingly, the report went nowhere. Whatever its conclusions, the report was hardly a product of Train's commitment to conservation. At this stage of his life he simply did not give it much thought. He held no opinion of the Tidelands Oil debate, the contentious question of whether Washington or the states should own the oil-rich lands submerged immediately off the coasts and one of the major conservation issues of the day. He had nothing to do with the many battles then taking place over the disposition of public lands. These were debates for others in government, not for him. At this point the Internal Revenue Code was his domain, the page after page of minutiae now forcing him to wear reading glasses.[47]

The work was worth it, providing Train a tutorial on Capitol Hill politics. The necessity for compromise, the give and take necessary for legislation, and the manner in which the committee members worked together suited him. In a sense it was easy; the Republican and Democratic members were fiscally conservative. The Senate Finance Committee, for example, included Democrats Harry Byrd of Virginia, Tom Connally of Texas, and Robert Kerr of Oklahoma, all southerners hardly known for their liberalism. Committee meetings were rarely rancorous, members remaining friendly and even socializing together.[48] For Train, it all solidified his inclination for bipartisanship.

The job provided other advantages, expanding Train's circle of powerful acquaintances. Names such as Wilbur Mills of Arkansas and John Dingell of Michigan, the ranking Democrats on the Ways and Means Committee, may have been familiar to most only through the newspapers or radio, but to Train they were daily associates. Mills's knowledge of tax matters surpassed that of all other members, one of the few Train knew that had a thorough knowledge of social security legislation. His political demise years later, the product of an embarrassing incident involving the stripper Fanny Fox and a car driven into Washington's Tidal Basin, became fodder for late-night comedians but distressed Train, who appreciated more than most his many years of commendable service. In later years Dingell became known as an opponent of environmentalists, but their years of working together and their mutual love of hunting always kept him and Train on friendly terms. Train also worked closely with Louisiana Democrat Hale Boggs, who later died in a tragic plane crash in Alaska, cutting short a career that had him poised to replace Texas Democrat Sam Rayburn as Speaker of the House. Train's friendship with Boggs led to a friendship with his son, Tommy, a well-known lawyer in Washington. Republicans of note included Robert Winthrop Kean of New Jersey, whose son later became a popular governor of that state. To members such as these, Train became known as a "tax philosopher," a term coined in admiration. "I never knew what a tax philosopher was," Tennessee Senator Howard Baker later wrote, "but whatever it is, I am sure Russ Train qualifies."[49] Other notables Train met included individuals with whom he would serve in the future, for example Elliot Richardson, later Nixon's Attorney General but then a fellow staffer on another congressional committee.[50]

Train was never shy. Once, sitting behind Senator Kerr as the Oklahoman made an emphatic point about a tax bill on the Senate floor, his voice resonating with the conviction for which he was known, Train interrupted him. "That's simply not true, Senator," he boldly stated. Glancing down at his young colleague, Kerr replied, "Train, when you are on the floor of the Senate you have to shoot from the hip."[51] Perhaps a shot from his own hip, but Train was confident enough to suggest his own revisions to the tax code. Now a frequent bettor at horse tracks with his friend Orme, he recognized that billions of dollars exchanged hands every year, most of it wagered with bookies in cash transactions that completely escaped taxation. The solution was a flat 10 percent tax on all such wagers, Train proposed, paid by the bookie. Train knew that the government required all income reported, illegal or not. It was a unique proposal, prompting colleagues to kid him about the expertise he exhibited. It was also a proposal that impressed Stam, who pushed it with committee members. In the end, Congress passed the proposal and the Supreme Court upheld it. Train attended the oral arguments wearing a loud checkered jacket, prompting Justice Felix Frankfurter to assume that he was one of the bookies with an interest in the case.[52]

Train was good at his job and when Republicans captured both houses of Congress in the 1952 election and swept Dwight D. Eisenhower into the Presidency, fate smiled on Train again. Gordon Grand decided to accept an offer to lead the Olin Mathieson Chemical Corporation, which opened up the position of clerk to the House Ways and Means Committee. The title was somewhat of a misnomer; the clerk essentially served as chief of staff, working closely with the chairman in an age when fewer subcommittees existed. Given the power of the Ways and Means Committee, the clerkship constituted one of the most influential staff positions in Congress. The election catapulted New York Republican Daniel Reed to the chairmanship, and Stam quickly convinced him of Train's qualifications. Before he could even lobby for the position, it was his. Given a salary of almost $15,000, which placed him in the top 5 percent of income earners, Train had a staff of nine working below him.[53]

Train was now a man of prominence in his own right. It fell to him to control the committee's agenda, which included scheduling and running hearings, serving as a liaison to various government agencies and committees, dealing with the press and voluminous mail, and, most notably, interacting with lobbyists. This latter task was new for him. He understood the need for interest groups to have their voices heard and, given the importance of the Ways and Means Committee, such interests spanned the economic spectrum. Trade associations, unions, farmers, professional groups, and corporations, large and small alike, all found their way to his door. It was a "turbulent life with lobbyists pounding on your door all the time," he recalled.[54] Train made it a point not to rebuff his solicitors but neither did he embrace them; the potential for corruption was obvious. He did not attend the frequent fancy receptions and even developed the habit of escaping to a local drugstore counter to eat his lunch in solitude and anonymity.[55]

Debates over tariffs, social security, and the other issues constantly before the committee sharpened Train's expertise. Soon he was "a tax wizard," in the words of one newspaper, the resource of choice for all questions on the Revenue Act of 1950, the Excess Profit Tax Act of the same year, the Revenue Act of 1951, and all the other mind-boggling assortment of statutes that constituted the labyrinthine American tax code.[56] He worked closely with Reed, which was not always easy. Having served in Congress before Train was born, Reed was an honest man but quite imposing and frequently stubborn. Like Stam before him, he found Train impressive and increasingly relied upon him. With the Republican accession, an effort grew to revise the tax code, culminating in the Internal Revenue Code of 1954, the first significant revision of the tax code since 1939. It was the "keystone" of the Republican program according to Eisenhower, and Train had a major role in its passage. He traveled the country, seeking the input of bar associations, tax institutes, and other interested parties while giving speeches as an advocate for change.[57]

Train worked well with his political opposition, augmenting his long list of Washington friends on both sides of the aisle. Nevertheless, his efforts working for the Ways and Means Committee solidified his Republican credentials. Roswell Perkins, Assistant Secretary of Health, Education, and Welfare in the Eisenhower administration, lauded Train's "dedication to and tireless efforts for the Republican cause." In addition to the 1954 tax code, Train "masterminded the whole process of the Administration's social security bill through the House Committee," even as Reed was ill or absent in Rome. The resultant Social Security Amendments of 1954 helped "establish firmly in the minds of the American people that the Republican Party was not going to turn back the clock."[58] Indeed, the majority of Republicans in the Eisenhower administration had no intention of harshly attacking the Social Security System. Like their leader Eisenhower, they were moderates in their view of government. It was easy for Train to contribute to the Republican cause, because he shared with the majority a philosophical alliance.[59]

Elections were important to Train, for he knew that he was subject to the ever-shifting winds of politics. When a Princeton friend, Republican W. Thatcher Longstreth, ran unsuccessfully for mayor of Philadelphia in 1955, Senator Hugh Scott organized a lunch for the candidate, who invited Train to accompany him. At the lunch was none other than Vice President Richard Nixon. Nixon did not impress Train, because Nixon was an obvious partisan, the counterweight to Eisenhower's "hidden hand."[60] Nixon knew his reputation. When Longstreth got the courage to ask Nixon to come to Philadelphia on his behalf, Nixon replied, "That's probably the worst thing I could do for you."[61] Nixon knew the political winds as well as Train, but undoubtedly neither anticipated their future association. In the years to come, changing national priorities—and politics—brought them together again, their efforts forever changing the way America dealt with her environment. For now, however, both had other objectives and neither gave their introduction much thought. Each went his own way.

2

From Courts to Conservation

Early in 1952, awash in a sea of congressional politics and taxes, Russell Train received an invitation from a friend he had made at Columbia, Joseph Spaulding. Spaulding had just married Helen Bowdoin, whose sister needed a date to a local party. Would Train accompany Aileen Bowdoin Travers, recently divorced and new to Washington? Yes, Train agreed, ignorant that his life would never be the same.

It was Aileen's first night in Washington and the date included dinner with her mother and stepfather, Edward Foley. Foley and Train hit it off immediately. Foley had served as Under Secretary of the Treasury during the Truman administration and, while a Democrat, shared both Train's affability and love of politics. Their conversation interested Train but hardly captured his full attention. It was the twenty-seven-year-old Aileen, with her brown hair, bright smile, and outgoing personality that caught Train's eye, an "absolute knockout."[1] At the party the two quickly escaped the crowd, retiring to the garden, where they talked until the early morning hours.

Two years later, on May 27, 1954, they married, appropriately enough at St. John's Church. The ceremony was small with only immediate family present, but the Foleys hosted a large reception afterwards at their house on Wyoming Avenue, a brick mansion formerly the home of Supreme Court Chief Justice Harlan Fiske Stone. Many of the connections Train had made attended, including his influential boss Congressman Daniel Reed. Aileen proved a strong ally in the environmental struggles that followed. She encouraged Train's activism and, in fact, became a noted environmental advocate in her own right. According to the *Washington Evening Star*, Aileen was the type of woman easy to characterize, "chic, neatly coiffed, carrying a name that is listed in all the right places." Nevertheless, to describe her as such was to tell only part of the story. She was a determined environmentalist, "a woman with an abiding interest in the world's health."[2] Aileen remained Train's most important confidant at each turn, an uncommon relationship in an age when a majority of spouses were content in what Meg Greenfield termed Washington's "wife culture." Most acted as "attractive but prim adornment-adjuncts to their husband's careers." They were "meant to be seen and not heard at certain office and social functions and to grace the platform at the departmental swearing-in or late-night campaign rally."[3] Aileen attended her share of ceremonies and social functions but hardly remained quiet. In some instances Train's inclination toward bipartisanship left him open to those less scrupulous. Aileen brought a level of cynicism and a degree of caution, pointing out those not to trust.[4] She appreciated her influence. "How lucky we are," Anne Richardson, wife of Train's colleague Elliot Richardson, once said to her. "Our husbands care about what we think."[5]

Aileen brought something else to the table—her own prominence, connections, and wealth. While many men in Washington might offer the obligatory chuckle about how they had married "above" themselves, in Train's case it was arguably true. Her family also had American roots back to the seventeenth century, her ancestors prospering in shipping and land speculation and emerging as one of the richest in the colony of Massachusetts, one serving as head of the legislature and another later serving as governor. Their wealth spawned Bowdoin College in Maine, which today remains a prestigious private institution. While some of her wealthy ancestors held loyalist sympathies, Aileen was a direct descendant of Alexander Hamilton. Later generations augmented the family wealth and frequently engaged in philanthropy. They became major financiers on Wall Street, close friends of the family of J. Pierpont Morgan, perhaps the nation's most famous and prosperous banker. They were true American blue bloods, the "natural aristocracy" to which Jefferson referred.[6]

Growing up, Aileen called three houses home. Her primary residence in Manhattan, located between Park and Madison avenues, was an impressive stone mansion. In summers and on weekends the family ventured to Oyster Bay, Long Island, the former home of Theodore Roosevelt and others of influence. There their home, situated on the water, had tennis courts and a boathouse. In the winters the family often visited the home of Aileen's grandmother in Montgomery, Alabama, a beautiful southern home now serving as the governor's mansion. Aileen received what one might term a classical education, common for aristocratic girls. She attended museums and the opera. She learned to ride horses, ski, sail, and play tennis, a love of the latter two later shared with Train. She also enjoyed bird-watching, which perhaps was where the roots of her own love of nature could be found. Her experience was not the norm. A noted ornithologist who sought access to the Oyster Bay estate, in agreement for such, tutored the young girl on her nature walks.[7]

Aileen's world was just as Republican as that of her future husband, although outside of Washington politics it no doubt played less of a role. Her father despised FDR and his regulations of the financial industry and he assumed that his progeny would always share his philosophy. In fact, Aileen remained a Republican throughout her life, although, like Train, she was less ideological and more open to change. When she was a teenager, her mother divorced and married Foley, an ardent Democrat who had cut his political teeth in Roosevelt's Public Works Administration. He became good friends with President Truman, the couple having moved to Washington after Aileen left home. When Aileen was twenty, she joined her parents on Truman's personal train as they traveled to the Army-Navy football game in Philadelphia. On another occasion, they joined Truman at the White House; throughout Truman remained a "very kind gentleman, not pretentious at all."[8]

She had her own connections, including a friendship with Jacqueline Bouvier, later to marry Massachusetts Senator John F. Kennedy. Their mothers were good friends, which led to Jackie joining Aileen on a month-long European trip to attend the 1953 coronation of Queen Elizabeth II. Sharing a cabin aboard the *S.S. United States*, the pair spent two weeks each in London and Paris. Aileen became as close to Jackie as one could. Jackie was a very private woman, Aileen recalled, always play-acting. "You never quite knew what she was thinking."[9] Upon their return, their future husbands were together waiting for them. It was Train's introduction to JFK, although it occurred only in passing and their discussion was only superficial. They did not share a political affiliation but they did share a love for their respective girlfriends. Within a year both couples married, Aileen serving as a bridesmaid in Jackie's well-publicized wedding.

Train not only gained a wife but also a family. With her came two daughters from her first marriage, Nancy and Emily, ages seven and four respectively. Aileen's family all welcomed Train, even her politically disparate father and stepfather. Given the troubles of their own alcoholic father, both little girls took to Train, who adopted them and raised them as his own. The new family settled into a rough stone home on Woodland Avenue in Washington. Train benefited from Aileen's wealth as her family contributed $30,000 cash toward the $80,000 purchase price. Train agreed to make payments on the remainder, the mortgage alone almost five times the average price for a new home.[10] The price was worth it, for the house was beautiful. Sunshine shone throughout the structure, with a garden behind it that included great ancient oaks from the original forest. Azaleas surrounded the garden's stone walk, composed of tiles from the historic Willard Hotel, one of Washington's most famous. The house remained the family's home until Train completed his government service. In this span of a quarter-century, Train enlarged the backyard terrace, added a swimming pool and later a cabin for guests.

And the family grew. In October 1955, Aileen gave birth to Charles Bowdoin Train, "Bowdy" to his family. In May 1959, a daughter, Errol Cuthbert Train, followed. Perhaps the product of his growing career or simply a man of his own generation, Train was not an active father when his children were small. He might give the occasional bottle but never changed diapers; a nurse was present for much of the mundane work. Train loved his children and, as they grew, his relationship with them grew as well. They learned to respect him and were more likely to go to him than Aileen with their teenage problems. It was always Train who initiated the most intimate discussions, the facts of life or other potentially embarrassing conversations, perhaps demonstrating the tact and diplomacy for which his career was famous.[11]

His life had changed, but so were the political winds shifting. Only five months after his marriage, the family still settling into their new house, the 1954 midterm elections cost the House Republicans their majority and Train his clerkship. When Reed became minority leader, Train moved one door down the hall to become minority advisor. It was more than just a change in offices, however; it was a clear demotion. His staff reduced to just one, Train no longer set the agenda. He was not angry and held no grudges, because he knew the rules of the Washington game. Still, ambitious as ever, he wanted change and a new challenge.

Fortune smiled on Train again. Word arrived that there were two vacancies on the U.S. Tax Court, a system of federal courts thirty years old established to lessen both the tax litigation burden on the district courts and the plaintiffs' onerous adjudication process. Under old English common law, citizens might have a jury trial over contested litigation but first had to pay the tax. Under the alternative tax court system, plaintiffs did not have to ante up the tax but neither were they entitled to a jury trial. A judge decided all cases, the President appointing and the Senate confirming each judge, just as in other federal courts. Their courts intended for convenience, tax judges conducted trial calendars throughout the country. In more populated areas the courts ran almost continuously, while in small towns they met only sporadically. In either case, the docket was heavy for the judges, who often spent their days on airplanes and their nights in hotels. It was not an easy life but it attracted Train. "I didn't see it as a stepping stone," Train recalled, downplaying his obvious ambition. "I grew up thinking a judge was a pretty good thing." It was, he insisted, simply a new chance to serve. "Somewhere I got that bee in my bonnet."[12]

The appointment, however, was not going to be easy. Train was only thirty-six years old, exceptionally young for any federal judgeship. Invariably politics surrounded the appointment, and while Train was clearly a Republican, he hailed from the District of Columbia, which was dominated by Democrats and had no viable local Republican organization. Others had worked on behalf of their own states' Republican agendas, in the process earning the respect of party elders. Train, however, had no such opportunity. There were no local Republican elected officials to speak on his behalf and no political reward for his appointment. "There is no question but that residency in the District is a very serious handicap in a matter of this sort," Train complained. "It is probably the most difficult obstacle that I have to overcome."[13] In addition, his brief career had never included a private tax practice, a qualification many considered obligatory. Even his mentor Congressman Reed expressed reservations that Train did not have sufficient experience.[14]

Clearly the situation required Train to utilize fully all the connections and diplomacy he could muster. He unleashed an impressive lobbying campaign. He contacted Edward Murdock, Chief Judge of the Tax Court, fully cognizant that Murdock, a fellow Princetonian, had received his own appointment when he was younger than Train. Murdock told Train privately that he backed his nomination, but his comments in confidence did not keep Train from letting Reed know of his support.[15] He worked to gain the support of almost every member of the Ways and Means Committee, an easier task because all knew of his critical and impressive work. They, in turn, lobbied administration figures such as Secretary of Treasury George Humphrey, the Treasury traditionally with the largest say in potential appointments. "I can vouch for his high character and the conscientious manner in which he approaches all the tasks assigned to him," Congressman John Byrnes wrote Humphrey. "Mr. Train has participated intimately in the writing of our tax law."[16] Added Howard Baker, "I know of no one who has a better knowledge of Congressional intent than he does."[17] Train's advocates even contacted the White House, prompting a response from Assistant to the President Sherman Adams. Train, Adams promised, "would be earnestly considered."[18]

Train persisted, writing old colleague Gordon Grand and prominent lawyers in private practice. He solicited the aid of Roswell Perkins, who lobbied
on Train's behalf with the Republican National Committee and the Citizens for
Eisenhower Congressional Committee, in each instance painting Train as an
unwavering champion of the Republican Party. Train, in turn, used such lobbying with others to allay the concerns of those more partisan minded.[19] Allies
argued that the position's rigorous travel schedule required the stamina of a
young man and that the Eisenhower administration, with its grandfatherly, gray-
haired image, might gain politically from such an appointment. Train, perhaps
cognizant of his own nice-guy reputation, warned that Senate Democrats might
block the confirmation of one more partisan.[20] Using his family connections, he
solicited old friends at J. P. Morgan and Company. His new father-in-law, George
Bowdoin, corresponded with the Republican National Committee and his own
connections in Congress, one of the latter reminding the RNC that Bowdoin
"has been a substantial contributor to the Republican Party for many years."[21]
Charles Russell, now seventy-six years old, contacted his old acquaintances in
Washington on his son's behalf.[22]

It was an impressive campaign of an experienced Washington insider, but it
was only partially successful. H. Chapman Rose, Under Secretary of the Treasury and a friend to both Train and his family, proposed a potential compromise.
"Chappie," as Train knew him, had obviously proved his most important advocate. Given persistent concerns over his inexperience, Train might come to the
Treasury to head the Legal Advisory Staff, with a promise of a judgeship the next
time a vacancy opened. It was a good deal and Train knew it. Cognizant of the
advantages he enjoyed, Train contacted all who had lobbied on his behalf, letting
them know of his decision and thanking them for their efforts.[23]

The Legal Advisory Staff offered Train a position similar to the one he had
in his days on Capitol Hill. It reviewed all regulations proposed by the Internal
Revenue Service, worked on tax legislation, and represented the Treasury when
called before congressional committees. Working from his office on the west
side of the Treasury building overlooking the White House, Train once again
was awash in a sea of taxation and politics. "I signed a lot of regulations," he
recalled.[24] In his new position Train befriended Secretary Humphrey, bumping
into him in the elevator, eating lunch in his dining room, and even visiting his
residence, a large house on Foxhall Road—now the German Embassy. In one
instance, the socially conservative Humphrey inquired about a cast that Train
wore on his leg. He had broken his foot dancing the Charleston on his living

room carpet, Train explained. "And this is my top tax lawyer?" Train wondered if Humphrey thought.[25]

It was just over a year later, in 1957, when Humphrey called Train into his office. An unexpected vacancy had opened and, as promised, it was Train's if he wanted it. Train jumped at the chance, Aileen agreed, and a swift Eisenhower appointment and Senate confirmation followed. The appointment pleased the fifteen other judges on the court, with their overburdened dockets. If Train's youth and inexperience concerned them, they gave only the slightest indication. Chief Judge Murdock wrote Train of the possible staff he might employ but added, "My advice would be that you do not employ a law clerk immediately but start off by doing the work yourself, so that you will know all about it."[26] Moving into his office in the Internal Revenue Service building just off the Washington Mall, Train thus began a career as one of the youngest sitting federal judges.

He was a solid jurist. While most cases simply followed precedent, several required the judges to write opinions that broke new ground. In such cases, Murdock and the other judges would review the opinion before publishing it. Only once did the majority ever disagree with Train's opinions. Life evolved into a routine. When court was in session, Train increasingly handled evidentiary and procedural issues with confidence, occasionally calling short recesses for a cigarette break, a bad habit he later overcame. Traveling was not always easy. If he had a friend in the city, he might eat with him. Otherwise he ate alone, mindful that social interaction with the local bar might appear partial. It could be boring, although there were moments of color. One of his West Virginia cases involved a brothel accused of underpayment of taxes, evidence of which included the number of towels sent to the laundry. Having just checked into his hotel room, Train answered the door to find a comely blonde batting her eyes and asking if he wanted some "company." Train refused the obvious setup, shutting the door to a tirade of profanities.[27]

When in Washington, his prominence grew. He now served on the board of governors of Washington's Metropolitan Club, which was growing increasingly controversial for its refusal to admit women. He also served on the board of Saint Albans School, along with such notables as *Washington Post* publisher Phil Graham and John Warner, later a Republican Senator from Virginia. Train and Graham, who was almost a full generation older, shared little in common other than their status as Washington insiders. Their acquaintance, however, led the Trains to a strong friendship with Katharine Graham, who later succeeded her husband as publisher and wrote a Pulitzer prize–winning autobiography.

Train and Warner shared a political alliance, both viewing themselves as moderates trying to appeal to diverse groups and bridging partisan divides. Train's growing status warranted membership in the Committee of 100 on the Federal City. Founded in 1923, this group of prominent Washingtonians sought to ensure and advance the fundamental goals inherent in the original eighteenth-century L'Enfant plan for the city.[28] Active in his church, Train served on the vestry of St. John's and the executive committee of the Washington Cathedral.

Train enjoyed his status and did not take perceived slights well. Invited to a reception one Christmas at the Eisenhower White House, Train and the other federal judges had to wait in the basement while the politicians and dignitaries finished their dinner upstairs, "Rudolph the Red-Nosed Reindeer" played for their benefit while they waited. While others might have felt honored just to attend the White House, a peeved Train thought it belittling. When the Eisenhower administration gave way to the youthful Kennedys, Train found himself at a table near Robert Kennedy. Train, bold as ever, walked over to the new Attorney General, introduced himself, and attempted to strike up a conversation. He was aghast, however, when Bobby virtually ignored the introduction, believing his reaction an indication of superiority.[29]

Bobby Kennedy did little to warm Train to the new administration. Train admired the moderate approach of Eisenhower even as the majority of Train's generation relished the sight of a President "born in this century" and with young children—just like theirs—in the Oval Office. Although Train had known JFK for some time, their relationship never blossomed like that of their wives. Earlier, at a dinner party, Kennedy had made a disparaging remark about lawyers, a comment no doubt made in jest but one that offended Train, who was standing nearby. Nevertheless, the Kennedy administration did little to alter Train's daily routine and he continued traveling as his calendar dictated. When he was in San Antonio, Texas, one November, the Kennedys were also in town. Recognizing Aileen as their motorcade passed, Jackie Kennedy smiled and waved to her friend. It would be a sight both Trains would always remember. The next day the First Family flew to Dallas.[30]

Train had his routine down. While the requisite travel brought some change, he spent the majority of his time in his Washington chambers with his clerks. He did not have to spend every waking moment in his office logging billable hours like many of his friends in the large private firms nearby; he enjoyed time for other interests. Train and his family continued to travel, his tennis game improved, and hunting remained one of his diversions of choice. There remained one interest that surpassed the others, however, indeed one that seemed to combine the others—Africa.

Somehow, Africa was a life-long interest. At age eight, Train saw the movie *Simba,* made in East Africa. Several years later he read an account of three boy scouts who had gone on an exciting safari there. Africa seemed to the young Train the "mysterious dark continent," as one of his colleagues would later recall, the racial pun unintended. It was wild, exciting, and exotic. "We dreamed of exploration, of Livingstone, Stanley, Burton and Speke. We devoured tales of Shaka Zulu, Chinese Gordon versus the Mahdi, the Boer wars and the Battle for the Bundu. We identified with Cumming, Selous, Roosevelt and Hemingway. Africa had real men and real animals . . . elephants, rhinos, lions and buffalo. They were animals you could sink your teeth into. And in most cases they spoke English. It was our kind of place."[31] In later years Aileen recalled that when Train moved in with her, he did so with two suitcases of clothes, a black Labrador dog, and hundreds of books on hunting in Africa.[32]

Aileen probably embellished the story, but it did indeed tell of the awe in which Train held Africa—or, at least, the Africa of his imagination. Africans had served for centuries under white domination, the same stories that enthralled the young boy reflecting the racist assumptions of the day. Train was no racist but neither did the daily reports of the suffering of real Africa move him. It was the romanticized Africa that won his attention, an Africa in which colonialists represented civilization cast against the forces of the wild. Hunters were not elitist interlopers but admirable adventurers.

Just as his career on the tax court commenced, Africa was just as likely to be on Train's tongue as the tax code. Finally Aileen could take it no more and proposed that they take their own safari. Train knew immediately that he needed help. A friend, Charles Hook, recommended he see Post Master General Maurice Stans. Stans so loved his many African safaris that when Train visited him at his cavernous office, he told his secretary to cancel all of his calls. They talked for over an hour. The conversation was purely on logistics but further encouraged Train. Not only did it provide Train the connections he needed, it introduced him to a man with whom he would long serve. In the years that followed, both Train and Stans worked in the Nixon administration, often on opposite sides of environmental policy but always with a common love for their safaris. For now, however, the neophyte Train was ready to embark on one of the most important trips of his life.[33]

In mid-September 1956, Train and Aileen left for a month-long trip to Kenya. Arriving in Nairobi, they were soon off with the professional hunter Stans had recommended, George Barrington, and local trackers. Venturing into the wilds of the Northern Frontier District, it was all that Train had hoped—giraffes so dark they appeared almost black, rhinos in herds of over fifty, silver-back jackals on the side of the road, baboons, ostriches, kudu, warthogs, and a menagerie of exotic animals not exactly indigenous to downtown Washington. Hiking for miles, camping out with the sounds of lions at night, they hunted for meat and their ultimate trophy, the elephant. Wearing the traditional attire, including large safari hat, they learned the tricks of the trade: patient waiting, the importance of tracks and water holes, and the influence of wind shifts on animal migration and behavior. Aileen held her own, her face occasionally protected by white cream, her "warpaint" Train kidded.[34] Danger was real. A large rhino "in a bad mood" charged them and forced them up a tree. Later, when there was no tree to climb and another charged, Barrington fortunately prevented a mauling with one steady shot.[35] Exhilarated, their hunt for elephants successful, the Trains returned again to East Africa two years later for an even longer safari. Now traveling into the Congo and Uganda as well as Kenya, their efforts to bag the elusive bongo failed. Nevertheless, they left no less exhilarated. They had, for one, encountered gorillas with "throaty cooing and cheek slapping."[36] While they had endured torrential downpours with lightning striking nearby, its thunderous clap deafening, and large safari ants crawling into their tents, nothing diminished their love of the land, which grew with every passing day.

Once back in Washington, Train received an invitation from Stans to join his newly formed African Safari Club. This small group of wealthy hunters smitten with a love of Africa periodically met at Washington's Roma Restaurant for dinner, drinks, and tales of their exploits.[37] Train attended several of these gatherings but left with a nagging sense that there was a problem the club just seemed to ignore. Train, it turned out, experienced more than just the thrill of the hunt; he experienced the real Africa—not the romanticized version of his youth—and recognized its threat to all that his fellow club members held dear. In Kenya he encountered the poverty of much of its native black population, noting "women half naked . . . one with an enormous pile of faggots on her head."[38] Much of the country was "hideously overgrazed."[39] Perhaps more direct to his concern, his 1956 party stumbled upon poachers, illegal hunters who decimated animal populations with no concern for proper management and profit their only motive. Hearing Train's party, the five poachers fled their camp, leaving their dried illegal bounty hanging from poles and their fires still burning. Train and his colleagues had their permits and they knew that game wardens existed, but Train also sensed that the wardens were not up to the task. Once, they ran across several of the wardens' staff in the wild, which Train considered "very unusual."[40]

The guide Barrington explained the situation. Poaching was chronic and while most animals were not yet endangered, the future did not appear bright. Many of the natives barely survived; they were subsistence farmers or pastoral tribes. Some were nomadic but, regardless, conservation was not high on their agenda—survival was. They had little tolerance for animals that grazed in their fields and destroyed their crops. They were not likely to worry about extinction with empty stomachs. To make matters worse, their population was exploding in one of the highest birthrates in the world. Barrington was a typical colonial settler, of German descent with a low regard for the natives. Nevertheless, his comments resonated with the young tax judge.[41]

The more Train thought of the real Africa and its impact on the world he loved, the more complicated the problem appeared. The forces of independence had begun to sweep the land; decolonization was inevitable. What would happen when the Europeans pulled out? The English, and to a lesser extent the Germans, had cultivated allies in a minority of the population that might staff important administrative posts in any new government. They had, however, done almost nothing to prepare this group for management of their precious natural resources.[42] The days of colonial rule had spawned game reserves and national parks, but they remained the purview of whites only. While many of these whites lacked formal training, their love of nature mitigated their amateur status. Natives, on the other hand, had often faced restrictions on game hunting in the reserves. They had no reason to love nature and, logically, no incentive for future conservation. Train placed little faith in the native population, like many of his colleagues never entertaining the thought that millennia on the land might have provided the natives a certain wisdom to bring to the table, that some intrinsic love of nature might exist. No, a problem existed and it required outside intervention.[43]

Never one to rest easy, Train raised his hand in one of the African Safari Club meetings he had begun to find boring. The club, he proposed, should develop a conservation mission. They should do something before it was too late. The proposal met with a positive response and Train soon found himself the head of a committee to investigate. Among those on the committee were Stans and Kermit Roosevelt, appropriately enough the grandson of Theodore. Train had never been active in conservation before. He was not a member of any of the leading conservation organizations other than a passing association with Ducks Unlimited. He pressed ahead nevertheless, calling a March 1961 meeting of the committee in his court chambers. There the committee agreed to focus on training Africans, a tactic that exhibited a certain noblesse oblige. No one else was doing it and, with independence looming, the need was urgent.

Raising money was the first order of business. The tax expert Train knew immediately that the African Safari Club would not suffice; they needed a separate tax-exempt organization that allowed for deductible contributions. With the club's approval and the $15 dollar filing fee from his own wallet, Train incorporated the African Wildlife Leadership Foundation (AWLF), his first venture into conservation. Assuming the role of chairman of the board, Train did it all in his spare time, trying not to let his new avocation interfere with his vocation. He rented a postal box at the Benjamin Franklin Station across the street from the tax court and used his chambers as the AWLF office. In time he rented a small office in the National Press Building and hired a secretary to help handle the paperwork.[44]

Once again Train employed his connections. He had befriended Jack Block in Nairobi, director of Ker and Downey, one of the largest safari-outfitting companies in Africa, and convinced him to share his client list. He wrote every client, often using his traveling court calendar as a way to meet them personally. Train then contacted Paul Mellon's Old Dominion Foundation and won the AWLF's first grant, in large part due to his friendship with ODF director Monroe Bush. While the sums were still relatively small, the money raised allowed the nascent AWLF to begin its work. George Petrides, a friend and professor at Michigan State University, contacted Train about a young Kenyan, Perez Olindo, who wanted to make a career out of wildlife management. In this way Olindo became the AWLF's first beneficiary and major success, later becoming the first African director of Kenya's national parks. In doing so, he joined a long list of Train's loyal friends, always signing his letters to his mentor with "your African son." Employing the assistance of New York's African-American Institute, Train and his fellow board members established a selection and administrative process for other applicants. Soon a group of Africans were studying at prestigious American universities, all eventually returning to their native lands.[45]

Ultimately, Train knew, Africans would have to learn their trade in their own countries. When Bruce Kinloch, the chief game warden of Tanzania and another friend, contacted Train about his efforts to establish a school of wildlife management in an abandoned building in Mweka on the slopes of Mount Kilimanjaro, Train did not hesitate. Seed money from the AWLF followed, giving birth to the College of African Wildlife Management. In later years, the AWLF assisted in the establishment of a similar school for French-speaking Africans in Cameroon, as well as educational programs at a number of African national parks.[46] Not all welcomed the AWLF's efforts, however. Where the dying remnants of colonialism remained, radical whites still frowned upon any education for blacks. Once, promoting African wildlife management in South Africa, Train spoke to a group of whites only to have his comments met with complete silence. No one offered thanks or even responded. "They just went on with their business as if I had never said anything."[47] Perhaps exhibiting ignorance of a less malicious nature, some questioned Train about how a hunter could favor conservation. It seemed counterintuitive that one who enjoyed killing animals sought to save them. Even former diplomat Averell Harriman, a family friend, questioned Train about the perceived dichotomy. This sentiment arose periodically throughout Train's life, forcing him to explain the need for scientific management. At times Train might deflect questions with humor and self-deprecation. For example, in Dublin at a dinner the president of the Bank of Ireland hosted, a guest asked Train how he could "shoot a poor endangered species." Without batting an eye, Train responded, "When I shoot at them they are never endangered."[48]

Thanks largely to Train's efforts, the AWLF overcame such resistance and thrived. To promote its mission, Train published a bulletin, *African Wildlife News*. Employing the skills he had learned on the Saint Albans newspaper, he wrote all the copy and editorials, took many of the photographs, corrected the galleys, and even carried it all back and forth from the printers in the evenings. Contacting experienced professors whom he had encountered, Train helped organize a seminar at the University of California to explore additional educational alternatives. Chaired by A. Starker Leopold, a Berkeley professor and son of the famed conservationist Aldo Leopold, the conference helped expand support throughout academia. Soon the AWLF sponsored a regular six-week summer course at the University of Michigan in Ann Arbor. Through his connections with Secretary of Interior Stewart Udall, Train secured the assistance of National Park Service personnel in the program. Udall, a Democrat, impressed Train. "He had no reason to think that this was particularly important to his job," Train recalled, "but was very supportive anyway." Train equally impressed Udall, who agreed to sponsor with Kentucky Republican Senator John Sherman Cooper a reception and presentation on the AWLF's behalf at the Smithsonian Institute. When it came to the AWLF, no partisan boundaries existed.[49]

Balancing the AWLF and the tax court, Train received a phone call that resulted in additional demands on his time. One might have forgiven him if he had passed on this new opportunity, but his drive and ambition would not let him. Ira Gabrielson, president of the Wildlife Management Institute, asked Train to join him as one of the founding directors of a new American affiliate of the World Wildlife Fund (WWF). Not much older than the AWLF, WWF already had branches in a number of countries. Dedicated to "a massive attempt to raise money professionally on a world-wide scale and to feed it into conservation channels using the very best scientific advice available," WWF's goal was similar to the AWLF's.[50] Despite his already overburdened schedule, Train could not resist the larger stage.

Incorporated in Switzerland, WWF was fortunate to have two able leaders, Prince Philip, the Duke of Edinburgh, and Prince Bernhard of the Netherlands. Their efforts contributed to a budget that dwarfed the AWLF's and thus an affiliate in the United States was the next logical step. To launch the American affiliate of WWF, they held a banquet in June 1962 at the Waldorf-Astoria Hotel in New York. A large crowd attended and both princes spoke at length. The glitzy affair brought the desired publicity and money, and offered Train an entrée into the conservation elite. Well accustomed to socializing with the rich and powerful, Train launched a relationship with the two nobles that lasted for years.[51]

With its own contributions and membership continuing to rise, the AWLF budget approached almost a quarter of a million dollars within a few years. The staff expanded into a number of specialized positions and enjoyed new, more spacious offices on Connecticut Avenue. A director of publications produced several documentaries and a branch office even opened in Nairobi.[52] Within the community of the concerned, Train had become somewhat of a celebrity. Aileen endured the lack of free time with her husband but joined him on his more frequent trips to Africa and shared in his expanding circle of international friends. Train grew close to Ian Grimwood, the chief game warden of Kenya; John Owen, the director of the Tanzania National Parks; and Hugh Lamprey, the first director of the College of African Wildlife Management. A particularly close friend was William Eddy, who helped with grants for the AWLF but resisted Train's pleas for him to join the board. These and others often greeted the Trains upon their arrival in Africa, Train's trips now more business than pleasure. Appointments with government officials, educators, and African advocates left him less time for the hunting that had sparked his interest in the first place.[53] Once arriving in Dar es Salaam after midnight to learn that their hotel reservations had been canceled and no other accommodations were available, the Trains frantically called the renowned naturalist Jane Goodall, whose unlisted number they knew but who had no reason to know of their trip. "Ms. Goodall is away but she told me to expect you," a house boy explained in his lilting Indian English. Obviously word of Train's arrival had spread farther than he thought.[54]

Crisscrossing the continent to promote the AWLF invariably brought Train in contact with the local monarchs. After camping on the Awash River in one 1963 trip, Train met with the famed Ethiopian Emperor Haile Selassie to congratulate him on the establishment of a new wildlife service within his country. Escorted into the palace as several of the ruler's family members stood nearby, Train spoke to the emperor through a translator. After offering his congratulations, Train thoughtlessly noted that the new wildlife service consisted only of the emperor's relatives. The smiling Haile Selassie suddenly froze and a frown appeared as he heard the translation. Train knew instantly that he had violated protocol and struggled to recover. He noted that a Coptic cross that the emperor had given the National Cathedral years before still played a prominent role in festive occasions. In what seemed like hours, the translator repeated the comments and the broad smile returned. Train had saved the day, the incident only a momentary lapse in his diplomatic skills.[55]

Train took every opportunity to press his case. When the movie *Born Free*, based on the book of the same name, was ready to open, he used the occasion to stage a benefit. It made perfect sense. He knew the author of the book, Joy Adamson, and had met the principal character, the lioness Elsa. Employing his friendship with Alice Roosevelt Longworth, he arranged a meeting with First Lady Lady Bird Johnson. Train, Aileen, and Mrs. Longworth drove in the Longworth limousine to the White House gates only to be denied entrance. Mrs. Longworth, it appeared, carried no identification. Exasperated, the daughter of Theodore Roosevelt declared, "Young man, I grew up in this house!" Finally granted admittance, they met Mrs. Johnson in the Blue Room for publicity photographs and then ventured upstairs to the family quarters for tea. Suddenly President Lyndon Johnson appeared in his pajamas. Ordered to rest after a recent heart attack, he did not even think of donning a robe to meet the party. It was Train's first meeting with the new President, and it left him with a poor impression. The cultured Train recognized the First Lady as an ally in his cause but thought her Texan husband "crude."[56]

Train's reputation continued to grow within the conservation community. He now served as an officer of Theodore Roosevelt's Boone and Crockett Club, founded in 1887 and dedicated to "sport with rifle and conservation."[57] An honorary trustee of the Tanzania, Kenya, and Uganda national parks, he was an official trustee of the American Conservation Association and on the board of the American Committee for International Wildlife Protection.[58] Increasingly he gave speeches, not on tax law but on the threat to wildlife. He stressed the role of foundations, universities, and the public—not just the government— in stemming this threat.[59] More, it seemed, knew of Train as a conservationist than jurist.

One who knew Train was Fairfield Osborn. The elderly Osborn was one of the patriarchs of the conservation movement, a man before his time. The longtime head of the New York Zoological Society (NYZS), he had interests beyond the organization's chief endeavor, the Bronx Zoo. The son of a former director of the Museum of Natural History, Osborn had published in 1948 the seminal *Our Plundered Planet*.[60] He also had established an African wildlife conservation program under the auspices of the NYZS and, like Train, had served as one of the directors of the American WWF. Osborn welcomed the AWLF to his cause but worried that Train's fund-raising success threatened his own contributors, especially those in New York. Calling Train, Osborn noted the duplication of their efforts and asked Train either to join his organization or shut down the AWLF. Train's refusal and his retort that "we have lit a fire under you" brought a curt reply: "Train, you son-of-a-bitch!" It was an inauspicious start to a relationship that evolved to admiration and, ultimately, to a close friendship.[61]

Osborn, after all, had other interests. The same year that he published *Our Plundered Planet*, he founded the Conservation Foundation, a reflection of his broader concern for human population and the regulation of pesticides. Headquartered in an upper floor of an old building in Manhattan, a rickety cage elevator its grand entrance, the Conservation Foundation operated as a private think tank. Emphasizing the importance of expert research to address issues of resource management, it sought to influence government policies through conferences and publications.[62] Without capital funds, it was not a foundation in the classical sense. Rather it derived its budget from grants and philanthropy, in turn funneling that money to others with specialized interests or competence. It was, as Train later labeled it, "a philanthropic middle man," directing money intended broadly for charity into the narrow world of conservation.[63]

Its budget just under half a million dollars, the Conservation Foundation employed approximately six people in its New York office, most working on educational programs for elementary and high school students. Also on the staff, however, was the English scientist Frank Fraser Darling, a man of international reputation whom Train knew through his participation at the AWLF's Berkeley conference. The Conservation Foundation had a stellar reputation but it sought to grow, which led Osborn back to Train. In the time since their initial exchange, Train had participated in a number of NYZS functions, impressing Osborn with his intelligence and his energy. At the annual meeting of the NYZS, Train joined Osborn and other key supporters at a private dinner held at the Waldorf-Astoria. Train seemed to have all the Conservation Foundation needed: the official connections, the fund-raising expertise, and most importantly, a commitment to the cause.[64]

It all made sense. When Sam Ordway resigned as president of the Conservation Foundation in 1964, Osborn called Train to offer him the job. This was not a proposal to join a board of trustees, such as the several Train served on already. It was the chief executive officer, the CEO, a full-time job. Train would have to resign his judgeship and completely alter his life. It was a momentous and hard decision that Aileen summed up succinctly. "Russ," she said, "you are going to have to decide to be either a judge or a conservationist."[65] Joining Osborn for lunch at the University Club in New York, Train made his demands. First, he would need a salary equal to his judgeship. This was somewhat of a shock to Osborn, who was accustomed to gentlemen of privilege serving without compensation. Second, he would not move to New York. He was not about to abandon his beloved Washington, so central to his life and career. Negotiations naturally followed. In the end, Osborn agreed and Train accepted.[66]

Only one problem remained. Train did not want to resign his judgeship until he had completed all of his cases. These cases lingered for over a year, which was surely frustrating to Osborn, who needed his services immediately. Osborn endured, however, evidence of both his high regard for Train and the expectations he held. Finally, on July 29, 1965, Train submitted his resignation to President Lyndon Johnson, arranged by National Security Advisor McGeorge Bundy, another Train friend. "I am grateful to learn," Johnson replied, "that you will be applying your great gifts to the conservation effort—an endeavor which, more than ever, is crucial to the future of our nation."[67]

On Labor Day 1965, forty-five-year-old Russell Train assumed his duties as president of the Conservation Foundation. It was a new chapter in his life, a complete break that only a few years before would have seemed unbelievable. No longer was he a judge with an interest in wildlife. He was now a prominent conservationist dedicated fully to the cause. And as Train, Osborn, and all of his new colleagues surely knew, the task before him was immense.

3

The Conservation Foundation and the New Environmentalism

As Train began his duties at the Conservation Foundation, the sounds of folk singer Bob Dylan permeated the airwaves. "The Times They Are A-Changin'," sang Dylan, perhaps not fully cognizant of how true his lyrics would become. Americans had enjoyed almost two decades of unparalleled economic growth, emerging as the "affluent society," in the words of Harvard economist John Kenneth Galbraith.[1] They assumed without question the merits of urbanization and industrialization and enjoyed rising standards of living that promised the perpetual good life. Only now, however, were the costs of such prosperity becoming apparent. In 1962, a former researcher at the U.S. Fish and Wildlife Service and an acquaintance of Fairfield Osborn, Rachel Carson, published *Silent Spring*.[2] The book documented the impact of indiscriminate pesticide use but had ramifications far beyond the agricultural community. A best seller with over six hundred thousand copies sold in its first year, it challenged America's optimism in a way nothing had before. America had limits, the book implied; her lifestyle did have negative ramifications.

The limits were becoming apparent if one looked closely. A drive in the country might reveal miles of beer and soda cans strewn along the highway. Wrappers, cigarette cartons, and every form of litter imaginable added to the blight. Yards filled with abandoned automobiles. In place of scenic panoramas, one found tacky billboards advertising everything from soap to supper. The problem was obvious to First Lady Lady Bird Johnson, who helped organize a White House Conference on Natural Beauty just as Train took over at the Conservation Foundation. Her efforts, and those of Carson and others, were a first step to awaken America to the growing travails of her environment.[3] Too much of the good life apparently threatened the good life. Indeed, the times were changing.

Train already knew this. His efforts to save the African wildlife had taught him the importance of habitat. It was not enough to limit poachers. One needed to protect the watersheds upon which the animals depended. The animals' prey and grazing lands needed protection as well. Nature tied the fate of one species to another because each had its place on the food chain. In the end, one needed to contain human population growth, which exacerbated every threat. Conservation was more than what most Americans traditionally assumed; it was a concern for overall environmental protection. Like Carson, Train had adopted an ecological perspective. The solution lay in seeing nature in its entirety, including man, and treating it as a whole. It was for this reason that he joined the Conservation Foundation, a pioneer in a movement that was sure to grow.[4]

A broader perspective, however, brought risks. Calling attention to man's actions begged more conflict. Some advised caution. Laurance Rockefeller, a board member of the Conservation Foundation and one of its most generous supporters, worried that too great an emphasis on ecology might unnecessarily divorce environmental policy from human needs. Rockefeller recognized the threat that man posed and had cochaired the First Lady's White House conference.[5] Still, he questioned preservation for preservation's sake and related more to Progressive Era conservationists such as Gifford Pinchot than naturalists like John Muir. Rockefeller and Train soon became lifelong friends; any differences in their environmental perspectives were trivial compared to what they shared. Indeed, while Train had grown to appreciate the threat to environmental quality, his experience with the African Wildlife Leadership Foundation reflected the concerns of the traditionalists more than the emerging environmentalists. During his AWLF tenure, he had played little role in the initial legislative forays into air and water pollution and had done nothing to support the Wilderness Act of 1964, which provided for federal preserves largely free from human intervention. Even with his new ecological perspective, Train believed wilderness legislation had value only as part of a larger environmental policy that ensured balance and sufficient resource utilization.[6]

"Like politics," Train stated as he assumed his new role, "conservation is the art of the possible."[7] If he were to enhance environmental quality and ensure human needs, he needed to broach competing values and recognize the legitimate vested interests of others. Conservation meant preservation to many nature lovers. Like the transcendentalists a century before, they saw value, if not a reflection of God, in undisturbed nature. Resource managers viewed conservation as proper management, judiciously employing the bounty of nature while replenishing it whenever possible. The nation should not—and could not—stop growing. Many businesses assumed nature to be the foundation of prosperity and often worked to exploit it as rapidly as possible, thereby maximizing profits. For them, conservation often represented a threat to the economy. Train was emerging as an environmental advocate but sought common ground. He recognized a need for forceful federal action but a strain of states-rights Republicanism remained. "There is an appropriate and necessary role for the federal government in financing and developing coordinated programs that are interstate in character, and we must be honest in admitting that there is a need for federal initiative where states cannot or will not act," Train explained in one speech only months after leaving the tax court. "Still, federal programs must not be permitted to supplant local initiative and local responsibility."[8]

Not surprisingly, in the years that followed Train engendered criticism. While businesses often chafed under the environmental restrictions he implemented, environmentalists occasionally thought his advocacy tepid. In 1969, author John McPhee interviewed Train for his book *Encounters with the Archdruid*.[9] Responding to a question about David Brower, the charismatic leader of the Sierra Club and one of Train's occasional critics, Train remarked, "Thank God for David Brower. He makes it so easy for the rest of us to be reasonable." Despite his growing environmentalism, Train had never cultivated much contact with Brower, one of the more forceful advocates of environmentalism. Brower naturally replied. "Thank God for Russell Train," he retorted. "He makes it so easy for anyone to appear outrageous."[10]

Train had no formal training in ecology, although his growing advocacy required a crash course in environmental science. "It was quite a challenge," one friend recalled, "going from the value added tax to pneumoconiosis."[11] In his first week on the job, Train gave a speech in the Grand Teton National Park before a joint conference of the American Forestry Association and the National Council of State Garden Clubs. With Lady Bird Johnson present and with Wyoming Governor Clifford P. Hansen introducing him, Train explained at length that "natural beauty" required a broad view of conservation.[12] It was a theme he returned to frequently in the following years. "Conservation is directly related to the productivity of the soil and the water of the earth, a biologically healthy environment," Train stated in another speech several years later. Outlining the manifold threats from everything from nuclear power plants to the loss of open space, Train explained that no problems were separate or distinct. "They are simply among the more obvious aspects of the central problem of how man and his society can develop in creative harmony with the natural environment of which he is an integral part and on which he depends absolutely."[13]

Train took only a minor role in Lady Bird Johnson's White House Conference on Natural Beauty, but after assuming the presidency of the Conservation Foundation his role grew. He agreed to serve on a "citizen's committee" to follow up on the recommendations of the conference and in time incorporated it into his own organization.[14] Garnering more donations, Train moved the Conservation Foundation into modern and spacious offices on Connecticut Avenue in downtown Washington. While far from abandoning Osborn's pet concerns of overpopulation and pesticides, Train shifted the organization's emphasis from research and education to practical applications, always working to ensure environmental values in decision making. Train was now the lobbyist, using his Capitol Hill connections whenever possible. Numerous times he testified before Congress. "Evidence of environmental degradation is nowhere more clearly expressed than in the statistics showing the accelerating number of species extinguished by human action," Train stated to the Subcommittee on Merchant Marine and Fisheries in favor of a 1968 endangered species bill.[15] "Every modification of the physical environment brings about a series of changes, sometimes beneficial, sometimes adverse, often unforeseen," he added to the Senate Interior and Insular Affairs Committee the same year, once again returning to his new ecological perspective.[16] When he was not testifying, Train often wrote articles. "Conservation today has a new connotation," Train wrote for one education journal, undoubtedly wondering if his repetition of the point made him sound like a broken record.[17]

Under Train, the Conservation Foundation sponsored a number of conferences around the country. It held regional workshops on air quality control and published a regular newsletter. On the west coast of Florida, it funded a unique real estate development designed to protect heron rookeries and mangroves along the shoreline, which also protected fish populations. The project was an example of Train's philosophy. While prohibiting the development constituted the best alternative for the environment, Train assumed such a prohibition unrealistic. The resulting compromise was, nevertheless, better for the environment than the uncontrolled growth that might have taken place. Working to spread its message, the Conservation Foundation sponsored a growing list of environmentally conscious authors. Ian McHarg, a professor of landscape architecture at the University of Pennsylvania, wrote *Design with Nature,* which dovetailed nicely with the Florida development. Other books included Daniel Price's *The 99th Hour: The Population Crisis and the United States* and Anthony Netboy's *The Atlantic Salmon: A Vanishing Species?*[18] Train's friend Bill Eddy produced an unusual documentary entitled *The Big Squeeze,* which bluntly illustrated the growing threat to environmental quality.[19] The Conservation Foundation's staff expanded while Train made his case. In assuming his new job, Train did not abandon the AWLF. In fact, he continued to work tirelessly on its behalf. He was a busy man. The times were changing, but Train hoped to meet the challenge.

Life was good. His career and family prospering, Train enjoyed the life of the wealthy. He frequently escaped the hectic pace of Washington on his forty-foot wooden ketch, *Traveller.* During the sailing season from April to November, his family sailed down the Chesapeake Bay to the Eastern Shore of Maryland and Virginia. The Trains shared *Traveller* with family friends from Washington, alternating weekends but occasionally enjoying a joint cruise. With the sun setting, the anchor down, and the sails furled, it was time for Train's favorite drink, bourbon on the rocks with a lemon twist—the perfect antidote to the demands of the nation's capital.[20]

Such escapes only whetted Train's appetite for a more permanent retreat. Anchored one happy night in a scenic cove, Aileen suggested purchasing a farm along one of the nearby tributaries. Weeks later the Trains were proud owners of Grace Creek Farm, a 170-acre waterfront property in Talbot County on Maryland's Eastern Shore. The property fit the bill perfectly. Originally a working farm, the main house dated to the early eighteenth century. With ducks and geese, nesting ospreys, and oysters off their own pier, it was both beautiful and peaceful. With the help of a full-time gardener and maintenance worker, Aileen added to the beauty with a large flower garden kept in bloom six months of the year. Nearby stood fruit trees and a vegetable garden, which supplied the family with a bounty of various jams, sauces, and juices. The Trains added a large glass-enclosed sunroom facing the water but during cold weather still enjoyed the two wood-burning fireplaces in the original living room. A swimming pool and tennis court soon followed. After attempting to moor *Traveller* off the dock, Train sold the boat and came to rely upon the farm's guesthouse and relative proximity to Washington for entertaining. Over the years a long list of prominent Washingtonians ventured to the farm, no doubt enjoying the retreat from their hectic lives as much as Train.

Hunting remained a key pastime. The farm provided duck, goose, and quail hunting, and a few years later the Trains purchased a nearby 190-acre spread of land known as Lostock. Together their properties allowed for sharecropping. With several houses on their land, Train employed a number of tenants over the years who also assisted with the maintenance. The Trains now had a new set of acquaintances, not as prestigious as those in the nation's capital but important to them nevertheless.[21]

Train hardly confined his hunting to the Eastern Shore. In 1967, the same year he purchased Grace Creek Farm, Train inherited from his father-in-law George Bowdoin membership in the Long Point Company. Long Point, Canada, sat on the north shore of Lake Erie in a relatively isolated stretch of land opposite Erie, Pennsylvania. About twenty miles in length, it was essentially a large sand spit covered with forests and marshes and constituted an excellent waterfowl habitat. In 1866 a group of sportsmen purchased the land and maintained it as a haven for duck hunting. Regulating the hunting season of its members and digging ditches to improve water circulation in the marshes, the Long Point Company built a series of docks and buildings connected by a boardwalk. Members had their own cottages, while the company maintained common buildings to house the kitchen, dining room, and staff quarters. Over the years a great number of prominent individuals enjoyed the hunting, including King George V, who visited in 1883 while still the Prince of Wales.

Train shared his cottage with J. Pierpont Morgan III, who acquired his shares the same year as Train. Known as the "Sin Center" because it earlier had been the locale of the members' drinking, the cottage was all dark-stained chestnut, a large fireplace providing the only heat. A sofa and overstuffed easy chair sat before the fire. Outside lay the beautiful marsh, its colors constantly changing as the sun rose and set and its reeds and water hiding an impressive array of wildlife, from hawks to bald eagles. While conversation often revolved around politics, isolating Train as one of the more liberal of the company's wealthy clientele, Long Point was a great example of the conservation Train preached, successfully maintaining its property throughout the year.[22]

Long Point was an excellent retreat in season but Train sought a warmer locale for the winters. When her father died, Aileen and her two sisters inherited his property on Jupiter Island at Hobe Sound, Florida. Just north of West Palm Beach, Jupiter Island was one of Florida's earliest resort developments. The family of Nathaniel Reed, a Train friend and future Nixon administration colleague, founded the development, which offered members a club with tennis courts, a golf course, and a marina. With tropical foliage and manicured lawns, it was the perfect escape from the gray and wet Washington winters. Within several years Train purchased his own home facing the inland waterway. The back terrace of the ranch-style home overlooked a sloping lawn with thick vegetation providing the necessary privacy. Grapefruit, orange, lemon, and lime trees dotted the property. In later years Train added a two-bedroom guesthouse for his many visitors. Just as at Grace Creek Farm, a long list of Washington's elite soon arrived.[23]

Train's growing estate hardly diminished his love of travel, for him a vital component of the good life. In July 1966, his family joined him on one of his longest African jaunts, a six-week sojourn to East Africa that combined AWLF business with sightseeing and a safari. While Train conferred with Jane Goodall and renowned anthropologist Louis Leakey, the latter explaining his latest archeological discoveries, the children overcame a stomach virus to appreciate the African wilds that so entranced their father.[24] "It was a great chance for the kids to observe and see Africa at its best," Aileen recorded in her journal. "It looked like the cover of an AWLF brochure."[25] Once again, Africa provided thrills as well as beauty. As Train's family tried to photograph a group of elephants, one of the wild giants charged them, his head down and tusk curled. Aileen and daughter Nancy took refuge behind a tree but Train and daughter Emily frantically ran. The elephant, a large anthill fortunately deflecting it, gave up the chase and allowed the frightened pair to return to their group. Three days later a pack of wild hyenas invaded Train's camp. "Emily, get out of here," Nancy shouted, when she thought that one was in her tent.[26] If this were not enough excitement, the family joined Hugh Lamprey, director of the Serengeti Research Institute, as he piloted a small Cessna aircraft over the Serengeti National Park. Lamprey suggested that Eileen might enjoy making small movements with the plane's control yoke. Aileen suddenly pushed the controls forward, sending the plane into a steep dive. Faced with negative g-force and the sight of the African bush rapidly approaching, Lamprey was able to right the plane, but not before the family realized that they had narrowly escaped another potential tragedy.[27] In the end, the excursion may have lacked the tranquility of a trip to Long Point but it helped pass his enthusiasm for conservation on to the next generation.

If the next generation were to enjoy such adventures in the wild, Train believed, the nations of the world needed to coordinate their efforts. Many countries had areas whose natural values transcended the interest of any one country, areas whose heritage belonged to the entire world. A century before, the United States had created the first national parks because their assets were of value to citizens of all states. So should the world community, Train believed, identify and protect those areas of interest to all peoples. As president of the Conservation Foundation, Train participated in a White House Conference on International Cooperation, agreeing to serve on its Committee on Natural Resources.

There he met the committee's chair, Joseph L. Fisher, president of Resources for the Future, who raised the possibility of a "trust for world heritage." Train and Fisher discussed the idea and proposed to President Johnson a World Heritage Trust to include sites of both natural and cultural value.[28] It was a novel concept, rather optimistic given the bipolar Cold War world of the Johnson era. Preoccupied with Vietnam, the Great Society, and domestic unrest, the Johnson administration failed to act. This did not deter Train and Fisher, who drafted a proposal in 1966 for the International Union for the Conservation of Nature and Natural Resources (IUCN). The IUCN embraced the concept and together with the International Council on Monuments and Sites began work on specific provisions and on a possible convention. The following year, in yet another trip abroad, Train participated in the International Congress on Man and Nature, held in Amsterdam. Once again Train pressed his case. "I believe it will be particularly important at this Congress," he pleaded, "to launch an international cooperative effort that brings together in a unified program a common concern for both man's natural heritage and his cultural heritage."[29]

Train's lobbying did not immediately bear fruit but did plant the seeds for a concept that would grow in the future. Clearly Train tried not to let the good life he enjoyed dull his commitment to his cause. His vision for conservation grew along with his personal portfolio. To act now, he assumed, ensured the good life for others.

Train's good life seemed a world removed from the reality of most of America in the late 1960s. By the end of the decade a social revolution fully bloomed, launched in the former beatnik hangouts of New York's Greenwich Village and the hippie havens of San Francisco's Haight-Ashbury. Buffeted by affluence but paradoxically rejecting the materialism of modern society, the new youthful counterculture questioned traditional assumptions. The repression of African Americans and women, the apparent pointlessness of the war in Vietnam, and the drudgery of the average middle-class existence—all galvanized a potent cultural force that demanded fundamental change. The early 1960s planted the seeds for this revolution but events now radicalized much of the nation. The Beatles who once sang "I want to hold your hand" now belted out the lyrics "Why don't we do it in the road?" Train, entering middle age, the father of four children and the proud owner of three homes, found developments somewhat confusing. "Like most people my age I found it pretty difficult," he recalled. "The rules of behavior had been turned upside down." Too old for military service and with children too young, Train supported the Vietnam War but did not follow events closely and recognized his convictions far from firm. "I was simply not directly confronted with it," Train recalled, an amazing admission given his connections to Washington's political world.[30] While his children soon reflected the changing styles, they remained happy, well adjusted, and drug free, once again removing for Train a direct conflict with the new social order. The cultured and neatly coifed Train appeared to have little connection to, or in common with, the bearded and sandal-wearing hippies of the day. In fact, the prominent Washington insider, he seemed the personification of authority, the very object of youthful scorn.

Appearances were deceiving, however. As part of their critique of modern society, the young latched onto the growing environmentalism as a central tenet of their beliefs. By the late 1960s, the problem was more than simply trash that constituted an eyesore. It was now obvious to more than just visionaries such as Osborn, Carson, or Train that a crisis in environmental quality approached. In fact, the counterculture brought attention to the issue in a way that Train's many speeches in the preceding years could not. Environmental degradation— whether a messy oil spill on pristine beaches or mountains of rusted automobiles taller than nearby buildings—executives now realized, made great copy for television. The National Advertising Council released a poignant commercial that depicted an American Indian looking out over a littered landscape, a tear rolling down his cheek.[31] Train worked for publicity, but the changing national mores of the late 1960s truly brought it.

Train now had more allies in his cause. As he assumed the helm of the Conservation Foundation, only a quarter of respondents surveyed indicated that they thought air pollution a serious problem in their own community. For water quality, the figure was one-third. Two years later, during the "Summer of Love," half the respondents indicated a serious problem in both areas. By 1968 the figure was two-thirds.[32] Many of Train's conservation colleagues began to broaden their efforts, although some did not embrace the new course as readily as Train. When young, long-haired environmentalist Joe Browder, typical of the new generation of activists, arrived in Washington, most of the established conservation groups hardly welcomed him. Browder, it turned out, shared his last name with communist leader Earl Browder, prompting some to claim that he was part of a communist plot to infiltrate their groups.[33] With his moderate approach and Republican credentials, Train did not always share the activist approach of his new younger colleagues, his relationship with their elder statesman David Brower only one example. Nevertheless, he welcomed developments for the momentum they provided his cause. "The confrontational approach of Brower and others was not always my way," Train recalled, "but I recognized they had value for their publicity."[34] The Wilderness Society concurred: "We cannot suitably quarrel with youth who seek a new direction."[35] More united environmentalists than divided them, according to the Sierra Club's Philip Berry. "There was a lot of common action and agreement."[36] Even Brower, perhaps for a minute overlooking his criticism of Train, noted the new environmental movement as "quite united."[37]

A common enemy, the old saying goes, makes for strange bedfellows. From traditional conservationists to radical activists, the new environmental movement did have a lot to unite it. As Train worked in the Conservation Foundation, America's population neared two hundred million, its rate of growth approaching that of India. Across the nation's fruited plains sprang miles of mass-produced and prefabricated suburban homes. The suburbs grew six times faster than established cities, and as early as 1960 one-quarter of Americans lived in such homes. Real estate developers, financial institutions, utility companies, and landowners pressed for more intensive use of land, and local governments, anticipating more tax revenue from higher land values, complied.[38] Two-lane roads became four-lane roads and shopping centers grew in open fields. Because in many areas waste treatment facilities did not keep pace with greater population density, many communities dumped raw sewage into nearby rivers and lakes. This led to eutrophication, the overfertilization of water plants. The resulting algal growth blocked the sun from deeper plants, whose death and decay eliminated the remaining oxygen in the water. In time, the water would become devoid of all life, its ecosystem destroyed. With agricultural runoff, the dumping of industrial chemicals and municipal dredging augmenting the problem, the nation's waters appeared to be in a perilous state.[39]

On land the problem was trash. Municipal waste—residential, institutional, and commercial refuse—constituted millions of tons a year, the elimination of which created a municipal expense that only transportation and education surpassed. Many communities simply relied on open dumps, the least expensive alternative but one that contributed to disease and contaminated the nearby water tables. Rarely did municipalities allocate adequate funds for sanitary landfills or incinerators, the latter contaminating the air without appropriate pollution-control equipment.[40] Too often the nation's air reeked of an acrid yellow gas, the result of heavy industry burning oil, gas, and coal and thus emitting particulate matter and sulfur oxide. With automobiles contributing unhealthy levels of hydrocarbons, carbon monoxide, and nitrogen oxide, air pollution matched the problems found on land and in the water. Los Angeles residents coined a name for it—smog, an amalgam of smoke and fog.

Train did not wear the attire of the hippies, nor did his actions match the stringent nature of Brower and others but, like the youthful masses demanding change, he recognized the need for government action. Johnson's Great Society had begun to act, passing initial legislation to create air and water pollution standards.[41] In addition to the Wilderness Act, Johnson signed legislation to create a permanent fund for the purchase of parkland, to limit solid waste, and to create a wildlife refuge system.[42] Train's lobbying while at the Conservation Foundation helped but Johnson's Secretary of Interior, Stewart Udall, led the way, in many ways challenging the utilitarian biases of his predecessors. Environmentalists had, according to the Sierra Club, "no stauncher friend."[43] In Congress, Maine Democratic Senator Edmund Muskie took charge. Hailing from Rumford, Maine, a town so polluted that the stench from its timber industry permeated the daily life of its citizens, Muskie served as chair of the Subcommittee on Air and Water Pollution of the Senate Public Works Committee. As such, he was in a position to place his imprint on virtually every pollution bill before Congress.[44] The Democrats had cultivated a strong record of environmental activism but, perhaps reflecting the growing recognition of the problem, the issue was far from partisan. In the spirit of bipartisanship that Train valued and reflecting the Washington he knew, Republicans and Democrats worked together to craft a national response. Indeed, environmental attitudes reflected geographical differences more than party affiliation; concern grew everywhere but was strongest in New England and along the coasts.[45] In the arid mountain states of the West, where energy and extractive industries dominated, support was weaker. The Democrats had their environmental champions and so did the Republicans, who reminded Americans that the original champion of federal conservation legislation, Teddy Roosevelt, was one of their own. With leaders such as California Congressman Paul McCloskey, Train was not alone in his party.

The Johnson-era legislation was a strong start but Train recognized that much more was necessary. Given the bipartisan spirit of accomplishment and the rising tide of environmentalism flooding the nation, he was optimistic. Not all agreed with the goals of the new environmentalism and, even within the movement, differences existed. Still, Train thought that the time was ripe for a forceful federal response, a goal since his earliest days at the Conservation Foundation.

* * *

Given Train's prominence and his stature in the new environmental movement, government service inevitably called. When Congress created the National Water Commission (NWC), President Johnson appointed him vice chair. Led by Under Secretary of Interior Charles Luce, the NWC's task was daunting. Federal reclamation projects had magically transformed much of the western desert into productive farmland. They allowed for the explosive growth of cities in the postwar years, which depended upon a lifeline of sufficient water. Conflict was constant, of course, as various states, municipalities, and industries battled for their fair share of the precious resource.[46] The emergence of the new environmentalism only made matters more complex. The dams, canals, and irrigation networks upon which much of the population depended irrevocably altered the ecology of the area, diverting the natural flow of the rivers and polluting the waters. Johnson charged Luce, Train, and the other commissioners with defining a water policy equitable to all, no doubt glad to have the controversial issue at least temporarily off his desk.

The NWC undertook an exhaustive study. Train believed that his role was somewhat "symbolic," that his presence was "simply to show that environmental concerns were represented."[47] In reality, he proved an effective advocate for his cause, helping to reconcile the many competing interests. Despite representatives from the timber and agricultural industries, commissioners rarely argued. Train gave credit to Luce but the NWC's report reflected Train's influence. Noting that the economic analysis of water projects was unduly optimistic, the NWC recommended that the Bureau of Reclamation move away from large dams and ultimately reduce the number of federal projects.[48] Given the influence of western states, Congress did not act upon the NWC's report. The tumultuous events of the late 1960s overshadowed its work, which never garnered the public attention it deserved. Nevertheless, the report helped signal the beginning of the end for the era of massive reclamation projects.[49]

With his reputation continuing to grow, Train did not shy away from controversy. Perhaps because of his relationship with Luce, he agreed to serve on an Interior Department advisory committee to prepare for an international conference on the role of water in maintaining peace. He also agreed to serve on a National Academy of Sciences committee to study the Supersonic Transport (SST), a project as controversial as the largest dams. The sleek plane was truly a technological marvel, able to carry almost 300 passengers across the Atlantic in just two hours, traveling at an amazing 1,800 miles per hour. It stood to revolutionize the airline industry, if it were not for its attendant sonic boom and excessive engine noise. The Boeing Corporation, the plane's principal manufacturer, lobbied hard on the plane's behalf, noting its projected economic impact and competition from the Soviet Union. Environmentalists countered, however; citizen groups formed in opposition and joined such organizations as the National Audubon Society and the Sierra Club. Once again Train found himself across the table from vested interests in a contentious debate, his stature establishing him as an environmental advocate of choice.[50]

As Train delved into the minutiae of dams and planes, environmentalism continued to grow in Congress. Joining Muskie as a leader on the issue was Washington Democratic Senator Henry Jackson. Known as "Scoop" to Train and others, Jackson served as chairman of the Senate Committee on Interior and Insular Affairs. Recently the recipient of the John Muir Award, Jackson joined the Conservation Foundation and others seeking a way to ensure that all federal decisions incorporated environmental values. Jackson's committee staff became familiar with Wallace Bowman, a former Conservation Foundation employee who then worked for the Library of Congress's Congressional Legislative Reference Service. Bowman told them of the goals of his former employer. Recognizing that his own staff lacked the necessary expertise, Jackson contacted Train about the services of Lynton Keith Caldwell, a professor of political science at Indiana University who worked with the Conservation Foundation. Explaining that his committee lacked the funds to employ Caldwell directly, he asked Train if his organization would fund Caldwell's consultancy. Train agreed and thus Caldwell began work for the Senate committee.[51] It little mattered that Jackson was a Democrat and Train a Republican, for they both traveled among Washington's elite. "I always liked him," Train stated. "I saw him and his wife on the Washington circuit."[52] The two shared a mutual interest in ensuring that leaders did not forget the environment when they faced tough choices. In the end, these developments were fortunate. Caldwell proved the key architect of the environmental impact statement process so critical to the National Environmental Policy Act (NEPA). This legislation was then only on the drawing boards of Jackson and Caldwell. Soon, however, it would emerge in the Nixon administration and constitute one of Nixon's greatest environmental accomplishments.

The thought of any Nixon administration seemed fanciful only years before. After serving as Eisenhower's Vice President and narrowly losing the 1960 presidential election to John Kennedy, Nixon lost his 1962 bid to become governor of California. An angry Nixon snapped at the press, "You won't have Dick Nixon to kick around any more!" While pundits began writing his political obituary, Nixon labored behind the scenes to recover. Events worked to his advantage. Johnson's troubles with domestic turmoil and a deeply divided Democratic Party over the Vietnam War provided Nixon a road to rebirth. By 1968, when an exhausted and embittered Johnson declared that he would not seek reelection, Nixon astutely couched himself as the candidate of law and order and a man determined to bring "peace with honor."[53]

Train played little role in the 1968 election. While this may seem surprising given his environmental advocacy, the election revolved around the same issues that confounded the Johnson administration and diminished his own environmental record. In the wake of the Tet Offensive in Vietnam, the assassinations of Martin Luther King, Jr., and Bobby Kennedy, a series of racial riots, and the brawling disaster of the Democratic National Convention in Chicago, the nation was dangerously divided. Voters arguably recognized that environmental degradation posed the potential for mankind's ultimate doom, but months of turbulence seemed to offer that possibility immediately, framing the debate for Nixon and his Democratic opponent, Vice President Hubert Humphrey.[54] Humphrey might have exploited Johnson's strong record but raised the issue of the environment on only two occasions—the dedication of a Texas park in August and an Oregon dam in September. While his campaign literature virtually ignored the issue, Nixon mentioned it in only one of his eighteen "Nixon Speaks Out" radio addresses. Train might have bemoaned the environment's apparent lack of electoral clout but saw no significant difference between the two candidates, his political allegiance undoubtedly shading his view. Both parties' platforms included general statements in support of the cause and both candidates offered vague promises, but as an environmental advocate, Train knew Johnson's record better than most.[55] The environment, Train optimistically assumed, had always been an issue of bipartisanship and he would press the matter with whomever won. He had met Humphrey at a small dinner, just as he had once met Nixon. "I don't recall making any judgments at that time between Nixon and Humphrey," Train recalled.[56] He believed he could work with both. In the end, Train proved a good Republican. His vote went to Nixon.

Little did he know that an election would once again change his life. Within days Nixon asked Dr. Paul McCracken, later his chairman of the Council of Economic Advisors, to create a series of transitional task forces to study the full range of issues facing the new administration and make recommendations. McCracken quickly appointed more than two dozen groups to cover everything from taxes to social policies. Notably lacking, however, was a task force on the environment. His career focused largely on foreign policy, Nixon had no record of environmentalism and no personal interest in it.[57] In the hectic days just after the election, McCracken mistakenly overlooked the issue. Fortunately, however, his deputy was Henry Loomis, a trustee of the Conservation Foundation. Loomis reminded his superiors of the importance of the issue and recommended the necessary task force. When McCracken agreed, Loomis recommended Train for the chair. Train's connections and status as a Washington insider had once again worked to his advantage. Loomis had been a friend of Train's for years, serving on the Conservation Foundation's board at Train's suggestion. He knew of his friend's Republican credentials and his commitment to the cause they shared. When Loomis called with the offer, a surprised Train quickly accepted.

Train had carte blanche but little time. He quickly employed his knowledge of Washington and his many acquaintances to appoint a group of environmental advocates. Not surprisingly, he assured that its membership included both Republicans and Democrats, attempting to maintain the spirit of bipartisanship. Train later described the task force members as "diverse."[58] Indeed, he included members from science, industry, environmental organizations, and all levels of government. Either intentionally or because conservationists constituted the majority of his acquaintances, however, Train did not appoint any overt environmental opponents. The group included wise-use conservationists and environmental activists but no forceful critics. Among the members were such familiar figures as Joseph Fisher and Charles Luce. While several members hailed from California, only two members came from the mountain West, where extractive industries and grazing interests hampered the environmentalists' cause.[59]

Given this, the task force's deliberations proceeded smoothly. Meeting in total only once in an extended session, members quickly agreed that their job was not to produce an overall blueprint for the future but rather to emphasize what the Nixon administration needed to do immediately. "An important first step," Train recalled, "was to make sure what was already there was working."[60] New legislation would prove contentious and time consuming, members concluded, while existing programs were underfunded. The task force wanted to impress upon the new President both the importance of the issue and its urgency. They wanted to stress the role of executive leadership. Avoiding divisive rhetoric and pragmatic in its approach, the task force adopted its resolutions unanimously. Not one member offered a minority view or addendum.

The report was unequivocal. "We recommend that improved environmental management be made a principal objective of the new Administration," read the first recommendation. "While time is running out rapidly on our ability to arrest and hopefully reverse these trends, we now possess the knowledge and technology to begin the job. Do we possess the will?" The second recommendation was more specific: "We recommend that priority be given to improving the surroundings in which most of our people live their daily lives—in our growing urban regions." To accomplish this, "we recommend that emphasis be placed on performance, on making existing programs work." In addition, the task force recommended a revised bureaucracy: the appointment of a Special Assistant for Environmental Affairs; the expansion of the existing Johnson-era Council on Recreation and Natural Beauty into a Council on the Environment; and, finally, the establishment of an environmental coordinator within each federal agency. Any further reorganization "should be deferred pending . . . a National Commission on the Environment."[61]

Train and his colleagues knew their audience. After the bitter disputes of his predecessor's term, Nixon had campaigned on a theme of bringing the nation together. The report, therefore, described the environment as a "unity" issue, one that cut across social, economic, and geographical lines. Wise in the ways of Washington, Train sensed that political advantage motivated Nixon. Both the report and the cover letter that introduced it stressed the environment's breadth of support. An appendix with polls indicated public concern and published charts on the increasing number of state elections. Conveniently omitting that other issues had easily overshadowed the environment in the election only weeks before, task force members wrote that "the overwhelming majority of voters" were willing to support local and state bond referendums on environmental protection. In other words, their support was deep enough that they were willing to endure some economic cost. The environmental tide was rising, the report implied, and Nixon might rise along with it. To do nothing might leave the initiative to his political opponents. This was the first time Train had noted the environment as a political issue. He always preferred bipartisanship, of course, and he did not want to alienate his Democratic friends. Nevertheless, he saw Nixon as a "political animal" and believed that the merits of the issue would not alone sway him. He would do what it took.[62]

Only days after Train's task force issued its report, President-elect Nixon announced his cabinet selections. Many environmentalists did not welcome the news. For Secretary of Interior, Nixon chose Alaska governor Walter Hickel. In many respects, the appointment made sense. A self-made man, the forty-nine-year-old had served as western regional coordinator of Nixon's campaign. He hailed from a western state, the traditional locale of nominees, and as a governor was familiar with both natural resources and federal land issues. His gubernatorial record, however, was decidedly prodevelepmont. Under Hickel, the state of Alaska unsuccessfully challenged an Interior Department policy that halted all land withdrawals from the public domain until the federal government resolved native land claims and disputed oil rights. He opposed a strict nondegradation policy in regard to federal water pollution law and publicly encouraged oil development on Alaska's pristine North Slope. His investments in oil stock raised questions of conflict of interest and he was a close ally of James Watt, who was chair of the Natural Resources Committee of the United States Chamber of Commerce and an ardent opponent of strict environmental legislation. Watt's ill-advised comment decrying "conservation for conservation's sake" ensured the ire of both environmentalists and Democrats.[63]

As many Senate Democrats geared to block Hickel's confirmation, the Republican in Train was willing to give him the benefit of the doubt. "The Democrats selected Hickel as a target of opportunity," Train believed, overlooking the governor's questionable environmental record.[64] As Hickel promised his support for pollution legislation, Train sought to ensure that the nominee knew the importance of the issue and had the correct facts. Given the apparent partisanship in his nomination battle, Hickel was unlikely to turn to existing Interior Department employees, almost all Democrats. Train and his colleagues at the Conservation Foundation, therefore, put together a large loose-leaf binder of briefing materials, much of it focusing on the Interior Department's role in enforcing water pollution law.

After receiving the material, Hickel invited Train to his room at Washington's Wardman Park Hotel. With Hickel was his longtime aide from Alaska, Carl McMurray. Hickel was friendly but appeared a bundle of nerves, frantically walking around the room and speaking in short, quick sentences. Train found his inability to discuss substantive environmental issues frustrating, but later learned the real reason for the meeting. The President-elect's staff, most likely his newly appointed chief domestic advisor John Ehrlichman and aide Peter Flanigan, had recommended that Hickel select Train for Undersecretary of Interior. Hickel wanted to meet Train before making his decision. Such a nomination, after all, made perfect sense. Since Train was well known as an environmentalist, his nomination might pacify Hickel's critics. His record of bipartisanship was helpful in such a partisan debate. Finally, his many connections on Capitol Hill would prove useful.

After the meeting in late December, Train returned to Grace Creek Farm only to receive a call from McMurray on New Year's Eve. McMurray wanted Train to return to the Wardman Park immediately. Leaving his family and friends, Train chartered a flight to Washington. Hickel cut straight to the point. Would Train accept the position of Under Secretary of Interior? In later years Train denied that he hoped for an offer to join the new administration or that his Conservation Foundation briefing book was an announcement of his interest. Clearly the offer did not surprise him, however, for everyone anticipated the nomination as a counter to that of Hickel. Illinois Senator Everett Dirksen made that exact point on the Senate floor.[65]

Train promised an answer within forty-eight hours and returned to the Eastern Shore. He wanted the job but he also wanted the support of those he valued. Although Hickel had implied that his appointment was more than just political expediency, environmental window dressing, Train still worried over the specific responsibilities he would hold. Excited, he could not even hunt ducks with a friend as he had planned. The news thrilled Aileen and the family but Train sought as well the advice of close friends Laurance Rockefeller and Fairfield Osborn. Flying to New York, he met with Rockefeller, who naturally supported his environmental colleague. Certainly Osborn had more reason to object. After only four years at its helm, Train would leave Osborn's baby, the Conservation Foundation, without leadership. Osborn, in his mideighties and quite frail, could no longer take an active role. The blessings of his mentor were important to Train, who recalled that "I don't know what I would have done had he objected."[66] Fortunately for Train, such was not the case. Lunching at the Mayfair House on Park Avenue, Osborn applauded the move. "Let's drink to it," he declared, ordering dry martinis. "To the future!"[67]

After Train called Hickel to accept the position, he had to prepare to meet Nixon. Each of the transitional task forces was to brief the new administration at the President-elect's headquarters in New York's Hotel Pierre. Only a week before Nixon's inauguration, Train arrived to find a large room packed with people. A U-shaped table dominated the room, the most important figures seated, with subordinates standing along the walls. There Train met for the first time incoming Secretary of Agriculture Clifford Hardin. Hardin, whose new job included controlling the Forest Service and representing the interests of farmers, might have perceived a potential threat in the environmentalist Train. The two, however, got along well. Train also met Nixon's new science advisor, the renowned nuclear physicist Dr. Lee DuBridge. DuBridge remarked that his responsibilities would include the environment, a comment that Train might have taken as a jurisdictional challenge. Train, however, felt no threat and recognized a potential ally. Notably absent was Hickel, which several members of the task force took as an ominous sign. The nominee avoided the briefing, they complained, confirming what his critics charged. Having conferred with Hickel personally, Train did not share this view and accepted without question his explanation that his confirmation struggle allowed no time. Train led the discussion of the report, with his audience offering no objections and little reaction.

That night Train and the others joined Nixon at a banquet in the hotel's large ballroom to celebrate all the task forces. Train was a bit nervous, having learned that he was to sit directly next to Nixon and thus afforded a unique opportunity to lobby for his cause. Train did not take the seating arrangements as an indication of Nixon's interest in the environment but rather the "fine hand of Henry Loomis at work."[68] For hours he debated what to say. Arriving that evening, Train found the President-elect mingling, shaking hands, and greeting the more than five hundred guests. When the opportunity arose, Train introduced himself. Nixon smiled. "I think we have big plans for you," he stated, a comment sure to give anyone pause.

The task force chairs sat at a large head table raised on a dais. Nixon was in the middle with the rest of the guests seated at round tables throughout the room. Nixon appeared relaxed and gracious, the center of activity and animated in his discussions. Gaining his attention, Train admitted that he had long debated what to say. "I am afraid that is a hell of a way to approach a dinner conversation," he remarked. Nixon was nonplused: "That is exactly what you should have done." Having once again decided to impress upon Nixon the breadth of environmental support, Train then explained that environmentalism crossed all geographical and socioeconomic lines. His comments would, he hoped, appeal to Nixon's political instincts. Nixon politely listened and then responded with a comment that surprised Train: "I am sure you are right in the suburbs and among much of the middle class. But what about the blacks and the poor in the cities? What is the appeal of the issue to them?"[69]

In only a few seconds, Nixon had hit upon a key point. Just as Train's African guide, George Barrington, had explained more than a decade before, the poor, struggling for their existence, have more on their minds than environmental protection. Whether exhibited in a youthful counterculture or wealthy safari hunters, a strain of elitism existed in environmentalism. Train tried to respond, correctly noting that the urban lower classes suffered disproportionately the costs of environmental degradation. Nixon had, nevertheless, demonstrated his intelligence, knowledge of the facts, and keen analytical mind. He was impressive, in command of the room, and presidential in all respects. Nixon went on to deliver an excellent keynote address, not once employing notes. Train left the banquet that night unsure if his transitional task force or his brief debate with the President-elect had convinced the new government of the importance of environmental protection. Nixon, however, was clearly a man of talent and substance. Train did not know what the future held but was hopeful, ready to begin yet another chapter in his life.

4

The Department of Interior

On January 20, 1969, Richard Nixon took the oath of office as the nation's thirty-seventh chief executive. Standing before the American people, his comments indicated that despite his debate with Train only days before, the report of Train's task force resonated. For the first time in history, a new President included environmental protection in his inaugural address. "In rebuilding our cities and improving our rural areas; in protecting our environment and enhancing the quality of life; in all these and more, we will and must press urgently forward," Nixon declared.[1] The crowds cheered, the pundits applauded, and it appeared that Nixon genuinely cared about environmental quality.

The truth was somewhat different, as Train undoubtedly knew. Ever savvy to the winds of public opinion, Nixon accepted Train's argument that environmental advocacy paid political dividends. His immediate concerns were the Vietnam War and the economy, but Nixon recognized the environment as a way to cultivate a new constituency. Nixon's daughter Julie recalled that her father entered the White House aware that his predecessor "had been broken by the bitterness and the unrest."[2] Environmental protection, the new President assumed, appeared to be the perfect issue to unite Americans. He could silence his critics who anticipated a conservative reign of terror and preempt his opponents, who might try to coopt the growing environmentalism for themselves. Nixon did not personally give the environment a great deal of thought relative to other issues, but his administration could still use it to his political advantage.

It was a strategy difficult to implement. Hickel blundered in lobbying for his confirmation, taking James Watt along with him on a visit to Senator Muskie's office. Muskie exploded in anger and Nixon received adverse publicity. While the Senate ultimately confirmed Hickel, it delayed its vote and thus embarrassed the nascent administration as it began its work.[3] Train's confirmation as Under Secretary of Interior, however, was the perfect antidote. Train encountered only token opposition, with many senators affluent in their praise. "I know that he will serve in this position with distinction," remarked Idaho's Frank Church. "I think the country is very fortunate to have the services of Mr. Train," added Oregon's Mark Hatfield. Even South Dakota's George McGovern, who challenged Nixon four years later, praised the nomination. "I think of all the nominations that have been announced in the new administration," McGovern stated to Train, undoubtedly implying Hickel, "and there isn't anyone who has brought more personal gratitude than yours."[4] The press praised the nomination, which appeared to fulfill Nixon's inaugural comments.

Train's hearings were quite a counter to Hickel's. If senators politely raised reservations, they sought to ensure that Train was not too biased in favor of the environment, that his environmentalism would not preclude resource utilization or harm the economy. Church, for example, asked Train if he knew the importance of the Bureau of Reclamation to western prosperity. Utah's Frank Moss wanted to ensure that Train's desire to preserve the national parks did not preclude the enjoyment of the same. Western states dominated the Committee on Interior and Insular Affairs, and Train responded by explaining his moderate approach and faith in balance. With Train's many acquaintances on Capitol Hill, most committee members already knew that he was no radical.[5]

Train appeared the antithesis of Hickel. The Secretary of Interior was a self-made man who never attended college. Opinionated and blunt, he spoke his mind. He could occasionally misstep, exposing Watt to Muskie as only one example, and he had many enemies as well as friends. The Ivy League–educated Train brought all the refinement that Hickel apparently lacked. Thoughtful and diplomatic, he won support through compromise and persuasion. If the committee members and the public perceived Hickel as evidence of administration antipathy toward the environment, Train reassured them—just what the White House hoped. Train's appointment pleased many employees in the Interior Department as well. "Our qualms on the appointment of 'any hick will do' as our new boss at Interior were ameliorated when you came on board as Under Secretary," National Park Service Director George Hartzog wrote Train years later.[6]

Train learned of Hickel's bluntness early. Just as he settled into his new office but before his official confirmation vote, Train received a call from Hickel's assistant Carl McMurray. Hickel, McMurray explained, wanted Train to appoint Watt as Deputy Under Secretary of Interior. Train answered that he only had a handful of positions that he could appoint, while Hickel had hundreds. Train had not met Watt personally but, like most environmentalists, was aware of his reputation. No, Train concluded, he would not make the appointment. A few minutes passed and another call came. "The secretary orders you to appoint Watt as your deputy," McMurray declared, no less blunt than his boss. Train remained firm, perhaps thinking for a moment that his appointment was political window dressing after all. "It would have made me appear to be a patsy of some kind and destroyed my credibility as an environmentalist," Train recalled. "In that case," he calmly replied to McMurray, "I will carry out the order but ask that the White House immediately withdraw my nomination."[7] Train knew that he had leverage. To withdraw would reinforce the new administration's image as an environmental opponent, just what the White House hoped to avoid. It would make Nixon's inaugural comments appear hypocritical. If Nixon wanted to win the environmental vote, he needed Train. Hickel must have known this, because a third call soon came. "The secretary," McMurray stated, "says forget all about it." It was a complete surrender. In the end, Hickel learned something about Train. Train may have been diplomatic and open to compromise but was not easy to push around.

More than the Senate confirmation hearings underscored the importance of the environment and the difficulties of winning the political initiative. One week after his inauguration and one week before Train's final confirmation vote, crude oil from a Union Oil Company well off the coast of Santa Barbara, California, began to leak, forming a huge oil slick that winds spread for miles. Oil covered beaches renowned for their beauty and waves washed dead seals and other marine life ashore. Coastal birds struggled to survive in the sticky black tar. Television crews captured it all, including local citizens venting their anger. Just like Rachel Carson's book several years before, the Santa Barbara oil spill produced outrage and proved another impetus to the growth of environmentalism. It was, according to the Sierra Club, another "historic opportunity to dramatize the nation's need for a better and more livable environment."[8] The difference, of course, was that this tragedy happened on a new President's watch. It seemed an early test of Nixon's new commitment to the environment.

Train was the bearer of the bad news. Just before dawn a Department of Interior employee called Train at his Washington home. While the telephone woke Train, the caller, three hours behind, had clearly been up all night. Train immediately knew the news was bad but did not fully appreciate the event's significance. So as not to awaken Hickel, Train waited an hour to call him.[9] It was not a coincidence that Train was the first to learn of the spill. Department employees, already frantically working to limit the spread, recognized it as an environmental disaster. Like everyone else, they knew from the confirmation battles that Train was the administration's resident environmentalist.

Hickel reacted slowly, initially proposing a voluntary suspension of drilling in the channel. Train thought that this was insufficient and argued for a complete suspension, ultimately convincing Hickel. Train also advocated seeking "the advice of independent geologists and engineers."[10] Hickel once again agreed. When this group correctly recommended that the department open and pump key wells in order to relieve subterranean pressure and avoid a worse disaster, however, many in the environmental community did not understand the necessity and offered criticism. Certainly public skepticism was understandable, not only because of Hickel's Alaska record and confirmation hearings but also because they knew that the administration was under pressure from the oil industry. For one, Texas Congressman George Bush contacted Train on behalf of Humble Oil. Humble, Bush explained, "was concerned that their side of the picture be understood."[11] Moving to silence his critics, Hickel approved a permanent two-mile-wide buffer zone where no companies could drill and tougher departmental regulations, including additional casing on wells, required warning devices and more mandatory testing. Still critics persisted, demanding that the administration make its suspension on the remaining wells permanent. Although Train thought it wise to do so, Hickel hesitated. Such an action, he worried, might make the government liable to the companies, which had paid millions for the right to drill. If the government were to agree, environmentalists might push for the same elsewhere, a solution he believed untenable given the nation's dependence upon foreign oil. In the end, after letting tensions cool and the attention wane, the administration finally agreed, fortunately avoiding any significant negative ramifications.[12]

Democrats, however, saw an opportunity. Muskie called public hearings to investigate the disaster. Noting that he had been "trying for several years to legislate in this area," Muskie characterized the administration as indifferent.[13] California's Democratic Senator Alan Cranston called for a "moratorium on drilling in all other federal tidelands in California" as the Sierra Club compared drilling to "America's landscape ravaged in the last century."[14] It was obvious to all that the growth of environmentalism had ended the era of bipartisanship on the issue. If the confirmation battles did not make this plainly clear, the reaction to the Santa Barbara oil spill did. Train thought the administration slow to respond but, in the end, acted responsibly. He bemoaned the growing attack mentality even as he encouraged Nixon that environmentalism reaped political dividends. Train sensed, after all, that political competition might play to the environment's advantage.

Just over two weeks after Nixon's inauguration and as the Santa Barbara oil spill unfolded, Train took his oath of office from Supreme Court Justice Stewart Potter, another prominent family acquaintance. Potter had lived close to Train's parents and had been a longtime friend, giving the oath at Train's request. Standing behind Train was Hickel, the man on whom Train's effectiveness as an environmental bureaucrat largely depended. Most observers undoubtedly assumed that the two would immediately clash, if not because of their apparent disagreement over policy then because of their disparate personalities. In fact, however, both were ready to work together. Perhaps because of the criticism he endured, Hickel surprisingly emerged as an environmental advocate much like his subordinate. Hickel had great plans, it seemed. The national parks, he lobbied, offered an excellent way to demonstrate the administration's commitment to the environment. Conveniently ignoring that his recommendation expanded his own authority, Hickel proposed a program "to bring parks to the people." The parks were virtually under siege, congestion burdening them and, in a sense, making them victims of their own success. The nation should spend more money, add more parks, and better maintain those that it already had. The idea "elated" Nixon, Hickel recalled. Train could not have lobbied better.[15]

Hickel, never one for details, had his eye on the big picture. "I manage from the top down," he stated.[16] He focused on the grand scheme and the politics that surrounded it, delegating authority to others. Train, on the other hand, was a hands-on leader, always delving into the nuts and bolts of each issue. Given his newfound environmentalism, Hickel apparently thought Train a perfect fit. Train had a large office, close to Hickel and away from the other top administrators. When he began his work, Train found his desk "positively covered with documents," a result, he assumed, of keeping Johnson-era holdovers from wielding influence. He did not know if Hickel had them sent directly to him but soon learned that he enjoyed a great deal of autonomy on most issues. Train's duties included the departmental budget and all legislative matters, a responsibility that covered every bureau and office in the department and placed him firmly in the middle of policy considerations. The Department of Interior included programs relating to water pollution, fish and wildlife, mineral policy, Native American affairs, parks and other public lands, reclamation projects, and, as Santa Barbara illustrated, oil leasing. Train had his hand in it all. He prepared legislation and responded to congressional proposals. He made decisions and facilitated decisions. Through it all, Train hoped to keep Hickel informed with regular briefings.[17] Hickel took a broad view, which allowed Train, in many respects and on many issues, the daily administration of the department.

Train understood that he was a subordinate. While he had forced Hickel's hand on the appointment of Watt, he still understood that he was now part of a team largely committed to a common cause. Train appointed a solid staff, including Curtis "Buff" Bohlen, Boyd Gibbons, John Quarles, and Jack Horton. They were all diligent workers dedicated to environmental protection. Bohlen and Gibbons, for example, later worked for the World Wildlife Fund and Resources for the Future, respectively. Hickel, however, kept his hand in appointments when politics were in play. While Train appreciated the career staff in the department regardless of their political affiliation—an indication of the bipartisanship he still valued—Hickel occasionally demanded that subordinates with Democratic leanings resign. Such was the case when Hickel ordered Train to fire Edward Craft, a member of the professional staff who had worked for the department for years. The press covered the dismissal, embarrassing Train, who did not agree with it. Train complied without complaint, nevertheless, aware of his role and doing his part to get along with his boss.[18] After Train's refusal to hire Watt, Hickel employed his old acquaintance as a consultant, which the press soon learned. Asked if Watt were in a position to influence policy, Train responded as Hickel did. No, both Train and Hickel claimed, Watt did not affect policy.[19] In fact, however, Watt pressed Train to limit jurisdiction over oil companies.[20] Train did not heed Watt's advice but obviously recognized that he had to operate as part of a team.

While Hickel and Train comprised a surprising team in their environmental advocacy, Nixon sought to implement the recommendations of Train's transitional task force. In a move that did not require congressional approval, Nixon issued an executive order creating the Environmental Quality Council (EQC) and expanding Johnson's Citizen's Advisory Committee on Recreation and Natural Beauty into the Citizen's Advisory Committee on Environmental Quality (CACEQ). Science advisor DuBridge chaired the former, an interdepartmental group comprising various cabinet-level officials and including Hickel. Train's friend and colleague Laurance Rockefeller chaired the latter, which was composed of conservationists and prominent private citizens. Overlooking both was John Ehrlichman, Assistant to the President for Domestic Affairs and rapidly emerging as one of Nixon's closest advisors. Ehrlichman, like his boss, not only recognized the political value in environmentalism but also held a genuine interest. For years he worked as a land use attorney in Seattle and acknowledged the importance of proper planning. Train believed that he could work with Ehrlichman and remained on "friendly terms." In later years, Train employed an intern at Ehrlichman's suggestion and Ehrlichman attended several of Train's Christmas parties.[21]

It did not take long for Nixon to realize that EQC was ineffective. Each cabinet secretary sought to protect his own interests while currying favor with the White House. "Like most interagency groups," Train recalled, "it tended to proceed on the lowest common denominator and tried not to ruffle too many feathers." Nixon grew impatient as Democrats pressed their case. EQC, Jackson complained, did not have "the time and energy to provide the continuity of effort required."[22] In public Nixon denied any problems but in private he agreed with Jackson. EQC, Nixon told Ehrlichman, "had become bogged down with a shotgun approach to all environmental problems." While Nixon told DuBridge that he was "anxious that the Council come up with three or four hard programs," EQC still did not offer one legitimate initiative.[23]

With EQC floundering, the environmental initiative fell to two groups within the Executive Branch. First, Nixon appointed another task force, this one run from within the White House and charged with composing a complete package of specific legislative proposals. Advisor Egil Krogh chaired this new task force but quickly decided to leave, frustrated by the complexity of the issue. Forty-three-year-old former campaign advance man John Whitaker replaced him, a solid choice given his doctorate in geology and, like Ehrlichman, his genuine interest in the matter. On the other hand, Nixon had the ever-present environmental advice of Train and Hickel. A sense of competition might have developed between the two groups, especially since Train held the Interior post that Whitaker coveted. Nevertheless, Train and Whitaker got along well. Unlike Whitaker's task force, the Department of Interior had responsibility for implementing most of the existing programs on the environment. It focused as much on the present as the future. His days filled, Train remained largely ignorant of Whitaker's work, a fact he later acknowledged was "hard to believe now." Whitaker cared about the environment but felt his first allegiance to Nixon. Train, he recalled, "served another master."[24] In the years that followed, Train's responsibilities grew but he always knew that having Whitaker inside the White House helped his cause.

* * *

Train wasted no time settling into his new duties. "We will make a practice of having weekly meetings," he wrote senior staff. "Except when otherwise noted, the meetings will be at 8:15 AM each Monday morning." The ambitious new boss, subordinates learned, was not one for a relaxed or lazy bureaucratic schedule.[25] "I am concerned that the development of new policy regulations within the Department is not being properly coordinated among all the different bureaus and offices," he later wrote. Now, Train ordered, all staff would coordinate regulations through his office.[26]

Proper coordination was essential because the issues that Train helped adjudicate were as complex and controversial as the Santa Barbara oil leases. Foremost was the proposed Alaskan oil pipeline. With the nation strongly dependent upon foreign oil, leading producers discovered the precious resource in Alaska's arctic north. While the North Slope remained unpopulated, it was far from a useless, frozen tundra. It constituted the last pristine wilderness in the United States, a land of soaring eagles and roaming caribou, the last great stand for many environmentalists. Anticipating huge profits, major oil companies, including Atlantic Richfield, British Petroleum, and Exxon, formed a powerful consortium to exploit the new fields, the Trans Alaska Pipeline System (TAPS). Its plan was to build a submerged hot oil pipeline 789 miles southward to Valdez, the northernmost ice-free port in the nation. Crossing federal lands, TAPS needed the Department of Interior's approval, thrusting Train into the middle of what promised to be a classic environmental duel. Complicating matters were the claims of the Native Americans, who had argued for a decade that the state's creation robbed them of land. Now they argued for a larger share of the potential oil profits. The situation was complex enough that former Secretary of Interior Stewart Udall had issued a "land freeze" prohibiting any withdrawals from the public domain pending a resolution of the matter.

Hickel recognized immediately that his connections to Alaska raised suspicions and thus he designated Train the head of the North Slope Task Force (NSTF), composed of representatives of the National Park Service, Bureau of Land Management, U.S. Fish and Wildlife Service, U.S. Geological Survey, Bureau of Indian Affairs, Bureau of Commercial Fisheries, and Federal Water Pollution Control Administration. Each of these agencies in the Department of Interior had an interest in the pipeline, but none more than Train. "I would like to see us seize on this as a major opportunity for demonstrating how environmental values can be cranked into development programs," Train argued, reflecting the goals he advanced at the Conservation Foundation. "Here would be a good starting point for designing guidelines for development."[27] Train wanted an established system to determine environmental costs, not only in regard to the pipeline but all federal programs. "One of the important functions of this Department," Train reminded his staff, "is to preview various projects to be undertaken by the Federal Government or pursuant to its approval in order to determine whether such projects would adversely affect one or more environmental factors."[28] Train did not immediately reject the idea of any pipeline, as some environmentalists urged, but once again sought to ensure environmental protection through a process of moderation and compromise.

This process required careful study. With lobbyists pounding on his door, Train ventured north with his assistant Jack Horton, whose background in geology proved fortunate. After consulting with representatives of the state, the University of Alaska, and various industries, Train flew on a TAPS airplane over the route, noting both the beauty and the rugged nature of the terrain. Once in the Arctic Circle, Train visited one of the giant oil rigs. Standing on its platform, he watched the sun set over the white landscape, light snow blowing in the wind but not obscuring a group of snowy owls patiently hunting for lemmings.[29] While in Alaska, Train received a telegram from Hickel ordering him to expand the NSTF to include other agencies outside the Department of Interior, most notably the U.S. Coast Guard and the Army Corps of Engineers. The President, Hickel wrote, made it clear that he thought the department should lift the freeze and allow construction. "It is urgent that we consider new ways which we can explore and develop the oil reserves of northern Alaska," Nixon believed, "without destruction and with minimum disturbance."[30] Train was not against expanding the NSTF, because he believed that the new agencies had a stake in the pipeline as well. Not including all interested parties "could create complications in carrying out our task." Many environmentalists worried that an expansion weakened environmental considerations but Train sought to reassure them. "We do intend," Train wrote Thomas Kimball of the National Wildlife Federation, "to work very closely with representatives of the conservation organizations."[31]

Train meant what he said. "I would like to reiterate the desirability of frequent communications with conservationists," Train wrote Department of Interior employees in Alaska. "The burden," he reminded them, "is on those who wish to use the public domain to establish their entitlement to that use."[32] Train went on to suggest several environmentalists he thought particularly important. Additional trips to Alaska followed, one of which was necessary for public hearings. Critics abounded but the judge in Train kept him composed. On his last trip, he traveled on Glacier Bay in a small National Park Service boat. There he noticed a merchant ship throwing garbage into the water. Train followed the ship and picked up the trash, including a cardboard box identifying the shipping company. A stern letter followed. "The waters of Glacier Bay are entirely within the National Monument," Train scolded, implying the possibility of legal action. "The American people, including your customers, visit an area like Glacier Bay in the hope of inspiration from the grandeur of unspoiled natural scenery."[33]

In September, the NSTF issued a series of stipulations that TAPS would have to meet in order for the Department of Interior to lift the freeze and grant the necessary right-of-way. Train favored this, noting that "lifting the freeze is independent from, and commits us in no way to, action regarding the application." Actual permits for construction might follow at a later date, he believed.[34] The NSTF's stipulations included the revegetation of disturbed terrain, the protection of streambed and fish-spawning areas, the free movement of wildlife across the pipeline corridor, and a ban on the use of harmful chemicals.[35] Train and Horton noted the problem of placing a hot pipeline with temperatures well over 100 degrees Fahrenheit into permafrost. It was, Train argued, akin to laying a red-hot poker on a cake of ice. The permafrost would melt, destroying the environment and weakening the pipeline. The solution, he suggested, was to raise portions of the pipeline above ground, always assuring that any structures did not impede the passage of animals.

Many environmentalists protested that this was not enough and, indeed, problems remained. TAPS, while promising to meet the stipulations, had yet to do so. Equally troublesome, the issue of Native American claims remained unresolved and many in Congress expressed reservations. Train knew that threats still existed, most notably oil spills from tankers in Prince William Sound. The last thing the nation needed was another Santa Barbara oil disaster. The wheels of bureaucracy turned slowly, however, and a final resolution would have to wait. Train, for one, had other matters before him.

"I am presently working practically around the clock, including weekends," Train wrote a colleague in March 1969.[36] He advocated tougher coal mine health and safety legislation, once again appealing to political instincts to aid its passage. "I am concerned that the Administration will lose identification with it," Train wrote Hickel. "Despite misgivings we may have about some details of the legislation, it represents a real landmark and could be a real political asset."[37] The tactic worked. The White House supported a tougher bill, overruling the objections of Counselor to the President Arthur Burns. Train pressed for a presidential message on population and family planning, a longtime concern of the Conservation Foundation, and submitted a proposal to Assistant to the President for Urban Affairs Daniel Patrick Moynihan. Moynihan agreed, and although the final text differed from Train's draft, Nixon delivered the first presidential message on population growth ever given to Congress.[38] When Nixon appointed the Garwin Committee to investigate the SST, a committee named for its chair, the renowned physicist Richard Garwin, Train lobbied to join the group. Noting that he had worked on the National Academy of Sciences SST Committee, he wrote Carl McMurray that Nixon wanted a "representative from Interior because of its environmental interests."[39] When the press reported public criticism of an increase in the timber harvest, Train did not hesitate to wade into the controversial issue. With the Forest Service in the Department of Agriculture, Train wrote to the Bureau of Land Management to do what it could: "I recognize the pressures and the reasoning for an increase but it should not be achieved at the expense of other important values."[40]

Train drove himself as hard as his staff. He sought to ensure protection for golden eagles and encouraged better solid waste management, including recycling. Nixon in the end agreed and emphasized recycling in his proposals to Congress.[41] Just after Fairfield Osborn died, in September 1969, Train did more than eulogize his Conservation Foundation mentor. He ordered a ban on several herbicides that threatened public health.[42] Pesticides had been one of Osborn's foremost concerns, so the White House's decision to restrict DDT pleased Train. Train, of course, was not alone in his actions. Nevertheless, it was obvious to all that his appointment was more than just political window dressing.

With Whitaker's task force working, Hickel pressed Train for new legislative proposals. "The Secretary continues to be concerned over the lack of legislative initiatives on the part of the Department," Train wrote in passing the mandate on to the assistant secretaries. "Please assume personal responsibility for canvassing the bureaus and offices under your direction as well as your own office for all such proposals."[43] The nation's coasts, Train soon learned, offered an excellent opportunity. Almost three-fourths of the nation's population lived in coastal states, the Great Lakes included. This population had grown by almost 80 percent over the previous thirty years, 30 percent higher than the national rate. Where only sand dunes once stood, high-rise hotels and beach cottages now crowded. Beaches and wetlands were the endangered lands of America, the battlegrounds for environmentalists of tomorrow.[44]

Train convened an interagency task force within the Interior Department, ordering everyone to remain quiet. "I do not want word out to the Congressional Committees or other agencies that we are drafting legislation," he wrote.[45] The result was a smartly crafted proposal that addressed the issue without overtly threatening state authority, a concern of many of Train's Republican colleagues. More carrot than stick, it established a Department of Interior program to encourage the development and operation of state coastal zone plans. The department would pay 50 percent of the cost of the plan's development and, once approved, 50 percent of its annual operational costs. The proposal specified criteria necessary for approval, including protection of ecologically fragile areas, the plan's supersedure over any local zoning, and compliance with all relevant federal regulations. States were free to participate or not.[46]

Train knew from his days on Capitol Hill the need for secrecy. South Carolina Democratic Senator Ernest Hollings was considering a similar bill and might try to preempt the administration. Congressional committees might try to alter the bill in a battle of jurisdictional oversight. In the end, Train's concerns proved prescient. Two years later Nixon signed the Coastal Zone Management Act of 1972. In many ways the bill kept the structure that Train and his colleagues first proposed, although increasing the percentage that the federal grants covered. The final bill, however, designated the Department of Commerce as the program's coordinator. While the Department of Interior could veto the approval of any state's plan, the House Merchant Marine and Fisheries Committee insisted on changing jurisdiction. Train's old Safari Club friend Maurice Stans, now Secretary of Commerce, proved influential as well, countering the efforts of Train. The final bill was a victory for the environment, although not to the extent that Train had hoped.[47]

Jurisdictional competition did not stop with coastal zone legislation; Train was forced to use his diplomatic skills and legislative expertise in other areas as well. With so much focus on the nation's waters, Nixon appointed the Stratton Commission to study ways to unite the federal agencies that dealt with the matter. With one of his aides on the commission, Train proposed a unified bureaucracy under the Department of Interior. The Interior Department already held the Bureau of Commercial Fisheries, Train argued, and it made sense to unite the management of land and ocean resources. Train had lobbied for environmental reorganization while at the Conservation Foundation but now found considerable resistance to his plan. The Subcommittee on Oceanography of the House Merchant Marine and Fisheries Committee pushed for the creation of the National Oceanographic and Atmospheric Administration (NOAA), and Stans argued that NOAA should fall under Department of Commerce jurisdiction. Train advised Hickel to avoid "unnecessary friction with the Oceanography Subcommittee." The smart political move was to "be friendly, yet noncommittal." It was classic Train insider politics but, in the end, did not match the clout of Stans. Commerce won NOAA jurisdiction, Train concluded, because it was a smaller department and because Hickel's outspokenness had begun to wear on Nixon. "It was somewhat of a political issue," Train recalled. A unified bureaucracy to deal with the problems of the oceans was a positive development for the environment, yet once again not to the extent that Train had hoped.[48]

Stans was a formidable opponent but Train's political skills were hard to deny. When former Florida Senator George Smathers lobbied for construction permits to build a housing development on Marco Island on the west coast of Florida, Train politely noted the destruction of wetlands and its impact on the surrounding bald eagle nesting sites and heron rookeries. The result did not prohibit development, the best alternative for the environment, but—another example of Train's philosophy—included permits with stringent environmental conditions. When efforts began to reduce funding for the TEKTITE program, a pioneering effort to develop and test underwater technologies in a special habitat constructed in the Virgin Islands, Train emerged as its strongest supporter. Train not only noted TEKTITE's successes but astutely won Hickel's support by invoking the jurisdictional struggles with the Commerce Department then taking place. "This project will help develop Interior's own capability for managing and carrying out marine programs," Train noted.[49]

Train's political skills were most effective in lobbying for his department's budget. He wrote key members of Congress, his words always polite, moderate, and deferential. "I sincerely believe this is a reasonable and practical request," he wrote Senator Alan Bible, chair of the Subcommittee on Appropriations, "and I earnestly solicit your Committee's support."[50] Train ordered all departmental agencies to refer to the Interior Department in their press releases. "This is not just a matter of bureaucratic pride," he noted. "I think it important to the funding of this Department that it improve its identity in the public mind." To improve Interior's clout, Train recommended the assistant secretaries participate in the relevant congressional hearings. Clearly Train knew how Washington worked. Characteristically, he sought a more uniform and programmatic system for determining budget needs. "Within this Department," he wrote the assistant secretaries, "we consistently are confronted with the problem of justifying increases over the prior year rather than having program plans evaluated against what should be done."[51]

In one instance, with Hickel out of town, Train had the "frightening job of presenting our final budget appeal to the President in person." Nixon greeted Train in the Oval Office with Ehrlichman and Bureau of the Budget chairman Robert Mayo present. "Well, I guess you agree with all of Bob's decisions on your budget," Nixon joked. "Absolutely not," Train replied. Nixon returned to his seat, tilted back in his chair, and put his feet up on his desk. Train then began a thirty-minute defense of the Interior Department's budget proposals. Nixon was relaxed even if Train was not. When the discussion turned to the need to purchase more parkland, Nixon stated, "I know something about real estate because it's about the only thing I can invest in!" Cutting the meeting short, Nixon took Train across the room to show him the instruments the Apollo astronauts had recently used to pick up moon rocks. Train wished the group goodbye and left. Only later did he learn that the White House approved all of the budget requests for which he had so aggressively lobbied.[52]

Train continued to work hard to maintain and cultivate his connections. Florida Senator Edward Gurney, for example, thanked Train for attending a dinner in honor of the Florida Republican Executive Committee. "If I may ever be of any assistance to you," Gurney promised, "I want you to count on me." Alaska Republican Senator Ted Stevens expressed similar sentiments for another reception. "I want you to know that I deeply appreciate your attendance," Stevens wrote. When Senator Karl Mundt won the WWF's Gold Medal Award, Train was the first to write congratulations.[53] Train kept in touch with his environmental friends whether writing condolences for defeats or praise for victories. He continued on occasion to advise the Conservation Foundation, which was arguably a conflict of interest, always marking his advice "personal and confidential" so as not to engender criticism. When Charles Lindbergh announced that he planned to leave the WWF's board, Train implored him not to do so. "Your identification with this cause has been a tremendous boost," he wrote. Train knew what struck a chord with the American public. When the journal *National Wildlife* submitted questions for an article on Hickel, Train was quick with advice: "If asked what your favorite outdoor activities are, say boating, tramping in the woods or horseback riding, but not hunting." Hunting, Train knew from experience, was not always easy to explain.[54]

Train enjoyed the challenges at the Department of Interior. He increasingly relied on the able staff he assembled and he developed new friendships, particularly among a group of under secretaries across the various departments. His friendship with Elliot Richardson grew and he lunched regularly with such notables as Under Secretary of Defense David Packard, the cofounder with William Hewlett of the Hewlett-Packard Company. Train did not have the opportunity to travel abroad as much as he would have liked, but at least on one occasion his duties drew him to Japan for a trip he would never forget. Cabinet secretaries regularly visited with their counterparts in Japan but, to avoid having all of them in the air simultaneously, under secretaries frequently represented their respective departments. Such was the case when Train flew directly from Washington to Nagasaki on a converted government cargo plane. Immediately bused to Kyoto for a lavish dinner and reception, Train and the other Americans could barely keep their eyes open. At the end of the dinner, Japanese ambassador Nobuhiko Ushiba sang a song of friendship and welcome. Train, who used to play tennis with Ushiba in Washington, recognized that protocol called for a reciprocal song. He glanced around the room at his exhausted colleagues, noting Secretary of Agriculture Hardin with his head practically on the table. Only Stans, seated directly across the table, looked sufficiently alert. The two stood and belted out the song "A Bicycle Built for Two." It was not embarrassing, Train jokingly recalled. "Every American was sound asleep and no Japanese knew what we were saying."[55]

The trip allowed Train to sign a northern Pacific fisheries agreement and meet Emperor Hirohito and Empress Nagako at the Imperial Palace. Security was tight. An assailant had earlier tried to attack Secretary of State William Rogers at the airport and, therefore, everywhere the Americans went armed guards accompanied them. With Aileen at his side, Train and another couple decided to enjoy some of the local cuisine. Before they could enter the crowded restaurant, however, security with machine guns had to sweep through the building. As he left Japan, Train received a commemorative photograph album with an exquisite silk brocade cover. Upon returning to Washington, he dutifully reported back to Hickel. "I pointed out some of the serious conservation and environmental problems faced by the United States and Japan," Train wrote, "and complimented the Cabinet Committee on its wisdom in starting the bilateral program to tackle these problems."[56]

Train knew from his days with the AWLF and Conservation Foundation the need for international cooperation. While his new position did not afford ample time abroad, he still took every opportunity to promote wide cooperation. When presidential advisor Moynihan suggested that the North Atlantic Treaty Organization (NATO) might have a role to play, Train strongly supported him. NATO existed as a political and military alliance among the western democracies but Moynihan suggested adding a third dimension to promote social and environmental cooperation. It could encourage the establishment of international standards and the exchange of technology, knowledge, and experience. It might also help diminish anti-American sentiments in allied countries critical of the Vietnam War. Nixon agreed and on the twentieth anniversary of NATO's founding, Nixon proposed the NATO Committee on the Challenges of Modern Society (CCMS).[57] "The Department of Interior supports enthusiastically the President's initiative in this regard," Train wrote Moynihan. To encourage success, however, CCMS should "initially restrict its programs to manageable proportions." Once it had mastered the complexity of the matter, it "could add to the breadth and depth of its activities."[58] Given the many ongoing environmental debates, Train obviously spoke as the voice of experience.

With his love of travel, Train may have perceived his Japan trip as welcome respite. In fact, his job back at home grew ever more complex. Even as the Santa Barbara and Alaska issues simmered, another controversy brewed on the other side of the country. The Dade County Port Authority (DCPA) proposed to build a massive jetport west of Miami. With runways six miles long, it could handle over two hundred thousand commercial flights a year from the largest jets. The proposed thirty-nine-square-mile complex was large enough to include the major metropolitan airports of New York, Los Angeles, and Washington combined. The problem was the proposed location—just inside the Big Cypress National Preserve, whose sloughs and saw grass savannahs provided the vital water supply for the famous Everglades National Park just to its south. At the tip of mainland Florida, the Everglades National Park's 1.4 million acres comprised a unique and multifarious ecosystem, home to over twenty-two endangered species of fish and wildlife and an array of exotic tropical flora. Opponents of the jetport organized as the Everglades Coalition, an umbrella organization of twenty-one environmental groups, even as proponents mocked them. When critics noted that the jetport's noise would drive alligators away, the DCPA's director promised to buy all the alligators earmuffs. Environmentalists were, he claimed, "butterfly chasers" and "yellow bellied sapsuckers."[59]

While the Department of Transportation had the authority to grant the necessary permits, the White House appointed a joint Transportation-Interior task force to study the matter, with Train as the chair. Train recognized the need to document fully the negative environmental impact and enlisted the assistance of Luna B. Leopold, a U.S. Geological Survey scientist and another son of the environmentalist Aldo Leopold. Leopold's report had the intended effect. "They were on their way to approving it until we blew the whistle," Train recalled.[60] "Your help," he congratulated Leopold, "has been tremendously helpful in our consideration of this problem." Train and Hickel, armed with Leopold's report, convinced Ehrlichman to oppose the project. In the end, the Nixon administration was able to convince the DCPA to relent and Miami constructed its jetport elsewhere. It took time and was not easy. Train fielded complaints from numerous jetport advocates, including Florida congressman Claude Pepper, who described Train's position as "unreasonable." Train remained polite, however: "I told him we appreciate having his views."[61]

Train and Hickel agreed on the jetport but the debate lingered for weeks and began to expose tensions between the two, the very problems critics had predicted at the outset but that both had sought to avoid. Looking back, Train recognized that the pundits were correct. "There was just no way for me to avoid problems," he claimed, acknowledging the role of his own ego. While Train handled the necessary details, Hickel still sought control and credit. "I couldn't just keep running back and forth to his office given all that we had to do," Train noted. "Somebody had to make decisions in a timely manner." Being part of a team was, apparently, harder than he thought. A conflict was inevitable.[62]

Once Train submitted the Leopold report to Hickel, weeks passed before Hickel or the administration took action. Train waited, growing anxious by the day. When an Associated Press reporter asked Train about the status of the jetport, Train replied that he thought it was time the Department of Interior moved on the matter. The subsequent story outraged Hickel, who already sensed that Train was trying to win environmental credit. "He wanted to burnish his environmental image and make sure that I didn't run away with it," Train believed. An angry Hickel summoned Train to his office and exploded. "This shit has got to stop," he yelled. Train knew that Hickel "had the right to chew me out," but still thought the episode "sort of ridiculous." Clearly Train had hit a nerve. Months before, the *Wall Street Journal* had run an article on Hickel. While the article noted Hickel's surprising environmentalism, its subhead read "Leaning on a Powerful Aide." In many respects it painted Train as the true environmental hero, the one actually in charge. No one had interviewed Train for the article and he never read it. Hickel fumed, nevertheless, laying the foundation for the inevitable rift.[63]

With Train's comments on the jetport, Hickel thought it appropriate to do what he earlier had threatened, to remove from Train his legislative and budget responsibilities. The news shocked Train, who sat "fairly speechless" and left "unclear of the actual assignment of responsibilities." As Train recovered, anger set in, and he decided not to take the de facto demotion lying down. The next day he talked with Ehrlichman at the White House for over forty-five minutes. He did not want to leave the administration and embarrass Nixon, Train explained, but he could no longer work with Hickel. Ehrlichman did not want him to leave either, with Nixon trying to win the environmental vote, the leverage everyone knew Train enjoyed. Ehrlichman, therefore, offered an alternative. Noting his dissatisfaction with the progress of the Environmental Quality Council, he suggested Train replace DuBridge as EQC chair. Cognizant that EQC's problems were more than simply a failure of leadership, Train politely refused and offered his own alternative. Congress, he explained, was about to pass the National Environmental Policy Act (NEPA). A key provision of this legislation created a new advisory body in the Executive Branch, the Council on Environmental Quality (CEQ). He would like to become, Train boldly suggested, both the chairman of CEQ and executive director of EQC.[64]

Train was well aware of the progress of NEPA. Ever since Senator Jackson had summoned Caldwell from the Conservation Foundation, Congress had debated the matter. Jackson's bill proclaimed the protection of environmental quality to be the nation's policy. Given Caldwell's expertise and the competition for the environmental vote in Congress, the bill now included a requirement that all federal agencies produce an "environmental impact statement" before completing all large projects. The bill's proposed CEQ would advise the President, issue annual reports, and oversee the impact statement process. Complicating matters, however, was a water pollution bill that Senator Muskie proposed. This bill included a provision for an Office of Environmental Quality, a potential rival to CEQ. In many respects, Jackson and Muskie competed for the title of environmental champion. Muskie was "not too happy about Scoop Jackson cutting into his issue," one of his aides recalled. On top of all of this was the administration's existing EQC, which despite its problems the White House still insisted publicly was the only council necessary.[65]

Train had long favored Jackson's bill. Train had the Caldwell connection, and he had met with Jackson in his Senate offices while still the president of the Conservation Foundation. As the bill advanced, however, Congress began to hold hearings and the White House insisted that Train testify in opposition. Train was uncomfortable but did so aware that Jackson and his colleagues knew his personal opinions. "I could see some of the twinkle in their eye when I would say that I think this is bad legislation," Train remembered.[66] After his testimony, Train lobbied hard for the White House to shift its position and support NEPA. Its passage seemed inevitable and opposition would hamper efforts to win the environmental initiative, he argued. Support would "identify the Republican Party with concern for environmental quality."[67] Michigan Congressman John Dingell, he continued, "practically ordered me to see to it that our Department does not oppose [NEPA]."[68] Finally, he added, CEQ did not necessarily mean that EQC could no longer function. "The two councils are not mutually exclusive and could easily coexist."[69]

Train's argument was persuasive; by the time the bill reached debate in the House of Representatives, the White House supported NEPA. Train testified once again, this time reflecting his true feelings. Muskie acquiesced and agreed to support Jackson's bill, no small development, since it meant that Jackson's Senate committee, and not Muskie's subcommittee, would have oversight jurisdiction. With the proposal for CEQ dominating the debate over NEPA, few legislators realized the importance of the bill's impact statement requirement. Even Train, who had argued since his days at the Conservation Foundation for some established program to weigh environmental considerations, did not grasp its ramifications. In the years that followed, this forgotten provision of NEPA defined much of national environmental policy and, as such, much of Train's daily work.[70]

For now, however, the tumultuous decade of the 1960s neared its end and Train, frustrated and angered with Hickel, recognized NEPA's imminent passage as his way out. While he earlier had suggested that Luna Leopold would make an excellent CEQ chair if Congress passed NEPA, he now pressed Ehrlichman for his own appointment.[71] Ehrlichman was noncommittal and Train "sat tight." Before leaving on its Christmas break, Congress finally passed the monumental legislation. Still no word came, but the press continued with reports of tensions at the Department of Interior. Attending a White House reception, Train had a welcome opportunity to speak with Nixon. Nixon was quite animated and, when gesturing with his hands, inadvertently hit Train's son Bowdy. Train had no reason even to think that Nixon knew of his request and was not about to raise the matter. Surprisingly, however, Nixon gave a hint of what was to come. "Someone needs to take charge of environmental matters," Nixon volunteered, "and you are just the man." The comment "flabbergasted" Train but that was as far as the conversation went.[72] A few days later, on January 1, 1970, Nixon signed NEPA while at his San Clemente, California, home. He declared that the new decade "must be the years when America pays its debt to the past by reclaiming the purity of its air, its waters and our living environment."[73] Nixon said nothing of his earlier opposition but rather portrayed himself as the ultimate champion of environmental quality.

An anxious Train could wait no longer. He called Ehrlichman at San Clemente, who told Train that he would raise the matter with Nixon and talk with Hickel. When a week passed, Train once again called Ehrlichman, who responded that, although Hickel surprisingly still wanted Train at Interior, Nixon had decided that Train should be at CEQ. Still Train worried. Ehrlichman had not said that he would chair the new council and Train believed anything less a demotion. Aware of his leverage and not afraid to use it, Train let Ehrlichman know that he would accept nothing but the chairmanship. Finally, a few days later, Whitaker called with the news. Nixon had decided to appoint Train the new chair of CEQ.[74]

Train was ecstatic. A new decade dawned and a new challenge was before him. He had served only a year at the Department of Interior but no one could doubt his impact. The end of the year brought retrospectives in the press of the momentous 1960s but most Americans looked forward with optimism. Certainly Train had reason to do so. Despite the extent and complexity of the problems facing the natural world, Americans apparently had woken to the threat. They now demanded action. Although political expediency motivated him, Nixon had joined the environmental bandwagon. He had finally settled on Train as the man to lead the struggle. It appeared to be the culmination of Train's career.

Captain Charles Russell Train (*far left*), naval attaché to President Herbert Hoover (*first row, third from right*); to Hoover's right is Charles Lindbergh (Library of Congress).

Nine-year-old Russell hiking in the Adirondacks, 1929 (Library of Congress).

At Fort Sill, Oklahoma, 1943 (Library of Congress).

When the good
Lord made Russell
Train he threw the
mold away, there
is none better than
he. Sincerely
Daniel A. Reed

Daniel A. Reed, Chairman of the House Ways and Means Committee, 1954
(Library of Congress).

Aileen and Russell Train on 1956 Kenyan safari (Library of Congress).

Fairfield Osborn (Library of Congress).

Train meets Alice Roosevelt Longworth (*left*) and Lady Bird Johnson in the White House (Library of Congress).

Russell Train and Secretary Walter Hickel, 1969 (Library of Congress).

Grace Creek Farm, 1970 (Library of Congress).

Nixon signs CEQ's second annual report in the Oval Office, 1971. *Left to right:* Train, Nixon, Gordon MacDonald, and Robert Cahn (Library of Congress).

Train meets with Spanish Minister of Economic Development Lopez Rodo and
General Francisco Franco (Library of Congress).

Dear Mr. Train
With my warmest regards,
Eisaku Sato
October 1970   Prime Minister of Japan

Train meets with Japanese Prime Minister Sato, October 1970 (Library of Congress).

Train meets Pope John Paul II at the Vatican (Library of Congress).

Prince Juan Carlos, later King of Spain, meets Train in Madrid, April 22, 1972 (Library of Congress).

Stockholm Conference, June 1972. Shirley Temple Black is behind Train (Library of Congress).

Train in experimental auto with NATO Secretary General Luns (Library of Congress).

President Ford greets Train in Oval Office, January 6, 1977 (Library of Congress).

Train (back row, third from right) meets with advisors to President George Bush (back row, far right) in the White House (Library of Congress).

Train pins J. Paul Getty Prize medal on President Ronald Reagan (Library of Congress).

Russell and Aileen Train, South Sandwich Islands (Library of Congress).

5

CEQ at High Tide

The great nineteenth-century British actor Sir Herbert Beerbohm Tree once remarked, "The only man who wasn't spoiled by being lionized was Daniel."[1] If Tree were alive today, he might have added Russell Train. Nothing prepared Train for the glare of national publicity that surrounded his appointment as chairman of the Council on Environmental Quality. Those who made their living "inside the Washington beltway" and within the larger environmental community already knew Train well. The new CEQ, however, cast him for the first time as a figure of true national prominence. The announcement came on January 29, 1970, only days after Nixon stressed environmental protection in his State of the Union address. Train walked into the White House Roosevelt Room with Nixon, a room jammed with the press and known to staffers as the "fishbowl." After Nixon introduced and praised his new chairman, Train had to answer questions for over thirty minutes. The next morning his picture graced the front pages of the *Washington Post, New York Times,* and other major newspapers throughout the country. He had interviews on CBS and the popular NBC morning program the *Today Show.* The weeks that followed brought more in-depth coverage with long articles on his life and career in such magazines as *Time* and *Business Week.* Although Train complained that several photos showed him "with my eyes cast down as if asleep," the coverage was universally positive.[2] He was "Nixon's own conservationist," the "pollution philosopher," the "big train in the nation's cleanup," and simply the "environmental protector."[3]

Train, it seemed, had arrived just as environmental quality peaked in the nation's consciousness. Even as the press lionized him, the nation prepared for what organizers termed a "national teach-in on the environment"—the first Earth Day. The issue appeared so popular that pundits declared that "no clear opposition to environmentalism existed."[4] Nixon followed his State of the Union with the first environmental address to Congress, a detailed packet of thirty-seven specific legislative proposals that Whitaker's task force had worked on for almost a year.[5] Its work essentially completed, the task force disbanded, removing for Train one of the other focal points for environmentalism within the administration. Train's responsibilities now covered more than simply the Department of Interior and he was free to focus as much on the future as the present. Nixon looked to him to secure the environmental vote even as the Democrats scampered for advantage in an increasingly partisan environment. The issue was terribly complex and the stakes high. Screw up, Train knew, and he faced the lions.

Train's two colleagues on CEQ did little to deflect attention from the chairman. They were able advocates and solid appointments whom Train welcomed, but neither held his clout. Dr. Gordon MacDonald, a renowned geophysicist and member of the National Academy of Sciences, was Vice Chancellor at the University of California, Santa Barbara. He brought technical expertise and important connections to the scientific community. Robert Cahn was a Pulitzer prize–winning journalist with the *Christian Science Monitor,* known for his outstanding environmental coverage. Meeting soon after their appointment in Train's Interior Department office, the three adopted what Train referred to as the "strong chairman mode." This implied a less formal working environment that essentially promised Train more personal autonomy.[6] The three got along well and worked in close proximity but rarely met in formal session. Each had his individual responsibilities but no doubt existed over who dominated the direction of the council or its agenda.

The pressure was as obvious as the glare from the nation's media. Just after the announcement of their appointments, Train and his colleagues joined Ehrlichman in Whitaker's office. Ehrlichman made it clear that CEQ was part of the Executive Office of the President. Its job was to advise and assist Nixon, not to lobby or criticize him publicly. Train was now the administration's top environmentalist but, at least in this sense, was still part of a team. His primary liaison at the White House was Whitaker, who worked with Nixon's cabinet-level Domestic Council. Environmentalists expected more, of course, as did many in Congress. In his confirmation hearings before the Senate Interior and Insular Affairs Committee, Train found a reception even warmer than his hearings a year before. Committee chairman Jackson invited Muskie and both were effusive in their praise. "That wasn't a confirmation hearing," according to one witness, "it was a love-in."[7] Clearly the environment's two leading champions on Capitol Hill did not expect Train to muffle his advocacy for political expediency. Train agreed with Ehrlichman and noted his record at the Department of Interior; he would not rock the boat. At the same time, however, he agreed with Jackson and Muskie; his new position afforded an unparalleled opportunity to advance his agenda. He wanted the boat to move even if he promised not to rock it.

It was a difficult balance, as Train soon learned. Not long after his confirmation, the National Press Club invited the new CEQ members to a breakfast interview. In response to a question from the *Wall Street Journal*, Train noted that serious environmental questions remained in regard to the SST. A few days later on the national television show *Face the Nation*, he noted that a fleet of SSTs created tremendous noise and contributed to depletion of the ozone layer. If this were not enough, weeks later he repeated the charge testifying before Senator William Proxmire's Joint Economic Committee. Nixon was furious. The administration now supported funding for two prototypes and Train appeared to take a position opposed to his superiors, a violation of his promise to Ehrlichman. Two prototypes, Nixon correctly noted, would not cause such problems. Properly chastised, Train wrote Nixon apologizing, undoubtedly wiser for the incident.[8]

The expectations Train faced made his task more difficult. Soon after his appointment, he joined Nixon, White House Chief of Staff Robert Haldeman, and Ehrlichman in Chicago to meet the governors of the Great Lakes states. Nixon surprised Train by asking him to summarize the administration's proposal to amend the Clean Air Act, which Whitaker's task force, not Train, had formulated. As he was ignorant of the matter, Train began to stammer, but Ehrlichman saved the day by providing the requested details. Train, embarrassed, tried to make amends afterwards but an obviously annoyed Nixon had little patience. Train still had a lot to learn but the more he interacted with Nixon, the more he recognized that his initial perceptions were correct. Nixon wanted to win the environmental vote but cared little for the issue itself. The President delegated authority but expected results. The pressure on Train was as great as the glare from the media. While in Chicago, Nixon and Train visited a sewage treatment plant. Standing before a large tank of clear, treated water, an employee took a glass and pressured Nixon to take a drink. "I never drink before lunch," Nixon smartly replied. On the helicopter ride back, a glum Nixon shook his head and commented that the public would remember Lyndon Johnson for saving the giant redwoods but would remember him only for sewage treatment plants. Train, most likely still simmering from his own embarrassment, thought to himself that Nixon should be so lucky.[9]

A year before, Train had sought to improve the operations of the Department of Interior. Now his organizational skills faced a greater task—defining, structuring, and staffing an important federal agency from scratch. He moved quickly, requesting that several key "working groups" and subcommittees from Whitaker's task force report to CEQ. The Jackson-Muskie compromise the previous year, which allowed NEPA to pass Congress, defined Muskie's proposed Office of Environmental Quality as an advisory body to CEQ. Recognizing some ambiguity, Train sought to ensure that this relationship was firm and that he, as CEQ chair, would control both committees.[10] He also sought to coordinate his efforts with the Citizen's Advisory Committee on Environmental Quality, which his friend Laurance Rockefeller chaired and which existed prior to CEQ. Train wanted CACEQ officially designated as another CEQ advisory body but recognized that he had to tread carefully. Several newspapers and conservationists had touted Rockefeller for the CEQ chairmanship and he did not want to alienate his old friend. Train decided to meet with Rockefeller and, while both avoided any official designation, they agreed to work together.[11]

Correspondence came from every corner of government. Commerce Secretary Stans, for example, offered the assistance of his department even as he noted industry's concerns about the new council. The newly created California Environmental Policy Committee requested coordination with its counterparts on the national level. Robert Cahn reported to Train that several congressmen "seemed to have very little awareness of who we were or what this Council did."[12] Establishing CEQ, it turned out, was almost as complex as the environmental issues it covered. The attorney in Train knew that sound legal advice was necessary and thus he created an Advisory Legal Committee. Never forgetting the importance of the environmental community, he also invited a wide range of advocates to a conference at the Airlie House in Warrenton, Virginia. "My colleagues and I are anxious to elicit the views of environmental leaders such as yourself," read Train's letter of invitation.[13] Clearly Train had not forgotten what it took to succeed in Washington. "You and your senior staff knew how to deal with the bureaucracy," recalled MacDonald. "Government was being creative, stimulated by your deep interest."[14]

Foremost, of course, was a solid staff and sufficient budget. Train lobbied ardently for both. "In consideration of the scope, complexity, and volume of matters which are being referred to the Council, as well as the matters which the Council itself will be initiating," Train wrote the Bureau of the Budget (BOB), "I have very real concern that even the requested level of staffing will prove inadequate."[15] Such lobbying paid dividends. While the number of staff never reached the level of Train's requests, CEQ employed fifty-four people at its peak. Most important were the top positions, of course, which Train took care to fill with expertise. Chief of staff was Alvin Alm, a senior examiner at BOB for water pollution programs. Also joining CEQ was Boyd Gibbons, who moved with Train from the Department of Interior. "I don't think that there was ever such a uniformly splendid collection of talent in one place as you recruited to CEQ," one official told Train.[16] While Train did not worry that he hired several Democrats, the White House did. "Al Alm, while not a Republican," Train reminded Ehrlichman, "has been intimately involved in politically sensitive White House matters throughout the Administration." He and others would cause no problems. Ehrlichman agreed, giving Train the benefit of the doubt.[17] CEQ staffers appreciated Train's bipartisanship. "When the White House political types discovered that I was a Democrat and, even worse, had written a book to which Ed Muskie had contributed an introduction, they were furious," recalled Terry Davies. "Internal pressure was applied to Russ to withdraw the offer he had made me, but Russ made it a matter of principle or honor that irrelevant political considerations would not affect the staffing of CEQ."[18]

Train, his colleagues, and staff moved into their new offices on Jackson Place, just off Lafayette Square and a block from the White House. Although Nixon told the press that CEQ would have offices in the Executive Office Building, Haldeman soon informed Train that a lack of space prevented it. Train did not mind, believing a physical separation from the White House implied at least some level of independence. CEQ's building was part of several homes recently constructed to resemble the nineteenth-century townhouses that existed there earlier. Train loved his new office and determined that it stood on the exact spot of one of Teddy Roosevelt's former homes. Given Train's job, it seemed appropriate. In his office were several George Catlin paintings of the American West, on loan from the Smithsonian Institution. Out his window he could see both the White House and Saint John's Church. Immediately behind his desk was an oil painting of a stern-looking orangutan, which he had enjoyed since his days at the Conservation Foundation. During one press conference, an Associated Press photographer took a picture of a scowling Train standing before the painting. The next morning the photo ran in all the papers, presenting Train with a facial expression similar to that of the hairy beast. Given his commitment to the environment, it too seemed appropriate. Train laughed off the unflattering photograph, one small example of both the positive and negative that came with the new national spotlight.

* * *

CEQ's creation hardly solved the problem of a muddled bureaucracy. Over eighty federal agencies dealt with pollution alone, an intolerable situation that assured jurisdictional overlap and dispute. With the dawn of the new decade, Nixon appointed the President's Advisory Council on Executive Reorganization to help solve the problem. This council, known as the Ash Council for its chairman, Litton Industries CEO Roy Ash, soon recommended expanding the Department of Interior into a Department of Natural Resources. Train thought the idea had merit but worried that the new department would also have responsibility for fighting pollution. "I believe such a department would be excessively unwieldy and exceedingly difficult to administer," he wrote Ehrlichman. To combine responsibilities for minerals and energy with environmental protection was to invite conflict. Returning to what he was sure resonated, Train stressed the political ramifications. Muskie would push for a separate pollution agency and, if the White House did not, he would portray Nixon as indifferent. The solution, Train insisted, was to advocate a separate Environmental Protection Agency (EPA), an agency others in the administration supported as well. Train and his allies lobbied the Ash Council and, in the end, won White House support.[19]

It was important, Train believed, that any EPA not infringe upon his own authority. The new agency should not have departmental status, he insisted. "Agency status would preserve for the President his freedom of choice in the future," diminishing the role of politics. The chairman of CEQ should serve as the head of EPA, he continued, in a blatant attempt to augment his own power. For those that might protest that a "three-man council could not effectively run the day-to-day operations of a line agency," Train responded that the chairman, not the entire CEQ, would have complete authority. CEQ's "environmental oversight function would technically extend to EPA" but, in reality, CEQ would remain the President's main advisory body, primarily responsible for the formation of overall policy. EPA, on the other hand, would serve as the main body for implementation of pollution law.[20]

In the end Train did not win control of EPA. "It was a little gambit I took in passing," he recalled, downplaying his obvious ambition. "No one thought it was a good idea." Nevertheless, one should not underestimate Train's role in EPA's creation. As CEQ chairman, he was the administration's chief lobbyist on Capitol Hill. He pressed senators for support and thanked them when they agreed. He testified on numerous occasions and, when problems arose, let the White House know. Congressman John Dingell, Train reported in one such instance, might resist. "From frequent remarks, it is obvious he is sensitive to the fact that he was not given credit for his role in [NEPA]," Train wrote. By the end of the year Congress passed EPA but not the Department of Natural Resources. Train claimed that he never wanted the job as the new EPA administrator, recognizing that environmental policy was still in its formative stages. Given Nixon's desire to win the environmental vote and the present state of legislation, he would enjoy more influence at CEQ than EPA. Train had never met the man Nixon did choose, Assistant Attorney General William Ruckelshaus, but learned of the appointment before the media did. A friend of Ruckelshaus, the ex-wife of Eisenhower's secretary of state, told Train of the news. Train called Ruckelshaus, congratulated him, and agreed to meet him for lunch. Like so many before him, Ruckelshaus became good friends with Train. "He never tried to fence me off in any way as a competitor in the administration," Train recalled. "At that first meeting, he told me that I would continue to handle international issues, which was very generous on his part."[21] Train clearly enjoyed the power and publicity but, in this instance at least, was willing to share both.

Train and Ruckelshaus agreed that the federal government needed "adequate environmental information systems that are readily useful to planners and decision-makers." Pressing Ehrlichman, Train argued that CEQ was too small to carry out any long-range analysis of environmental policy or to compile a data collection large enough for the use of all federal agencies and departments. Ehrlichman agreed and appointed Train to head a "working group to prepare a joint recommendation to the President in regard to a National Laboratory, Institute of Environmental Studies, and related matters."[22] Quickly contacting his connections in the scientific community, Train found interest high. The BOB also supported the concept, perceiving it as a way to subject environmental programs to stringent economic analysis. Existing institutions, Train argued, did not cover all the necessary scientific areas. In addition, many of those relevant lacked sufficient quality. By the end of the year, Train and his colleagues submitted a proposal for an Environmental Policy Institute. Nixon accepted the idea and officially proposed it to Congress in 1971. "I hope that this nonprofit institute will be supported not only by the Federal Government but also by private foundations," Nixon declared. "The Institute would conduct policy studies and analysis drawing upon the capabilities of our universities and experts in other sectors."[23]

Commerce Secretary Stans had his own ideas about government reorganization. Watching CEQ and EPA with some concern, Stans proposed a National Industrial Pollution Control Council (NIPCC), a group of industry leaders under

Commerce jurisdiction to "provide an effective liaison with the Federal Government." Industry was not the enemy, Stans argued, but a willing partner in the solution.[24] Whitaker, however, believed that he knew Nixon's priorities better. Nixon had just told him that the administration "needed to be harsh on industry to abate pollution" and to do otherwise allowed "Muskie and the Democrats to seize the initiative."[25] When Nixon agreed to the creation of NIPCC, Train and Whitaker decided to impress upon the new organization Nixon's commitment. NIPCC members would, in Whitaker's words, "be brought into really small, private meetings with Train," who would stress the importance of environmental protection. It was a message that Whitaker and Train planned to repeat as often as necessary.[26] Soon after the creation of NIPCC, which was already controversial for its refusal to admit representatives of the environmental community, Stans invited Nixon to attend one of its meetings. Train argued against Nixon's attendance. "Adverse publicity weighs against the President's attending the meeting," he wrote. Ehrlichman agreed and Nixon publicly avoided NIPCC, one small indication of Train's new clout.[27]

If any doubt about Train's clout still existed, Nixon quickly dispelled it. Soon after his environmental message to Congress, Nixon issued an executive order "for the protection and enhancement of environmental quality." The order solidified CEQ's primacy in NEPA's environmental impact statement (EIS) process, an action Train recommended. NEPA required that all federal agencies consult with CEQ in the creation of their statements. Nixon's new order, however, mandated that each agency comply with specific CEQ guidelines, including the requirement for public hearings. CEQ was to "coordinate" all federal environmental programs, monitoring agency compliance and enjoying a close working relationship with the White House.[28] "[Train] got an initial Executive Order written to give CEQ authority to issue guidelines for NEPA statements," CEQ general counsel Tim Atkeson recalled. "From then on, the NEPA process gathered steam in the courts and agencies."[29]

Train's advocacy came at a cost, however. In the fall of 1970, Whitaker asked him to brief the entire cabinet on the EIS process. Using a chart set on a tripod, Train used as his example the proposed five-megaton test of the Spartan antiballistic nuclear missile, scheduled for Amchitka in Alaska's Aleutian Islands. Train mistakenly attributed the test to the Department of Defense and not the Atomic Energy Commission. He also automatically assumed that the test was subject to the EIS process, a false conclusion given that the courts soon ruled matters of national security exempt. In any event, it was an unfortunate example given the Cold War and Nixon's commitment to a forceful foreign policy. As chair of CEQ,

Train had attended all cabinet meetings, sitting along the wall with other key agency heads and just behind the various secretaries at the table. Now, however, Haldeman informed him that he would no longer regularly attend. Train might have attributed the snub to a recent *Life* magazine article that described him as an eastern elitist, a group Nixon never admired. He knew, however, that he had no one to blame but himself. Train enjoyed the status attendant to cabinet meetings but rationalized that his absence in no way diminished his authority. "Cabinet meetings were pretty pro forma," he claimed. "It was not like I was missing any great action."[30]

Gaining agency compliance with the NEPA process was never easy; one arm of the bureaucracy always resisted oversight from another. "There is a built-in bias against anyone looking over your shoulder to see what you are doing," Train stated. "You have certain statutory responsibilities and you expect to get on with those without interference." It required a great deal of tact—fortunately a Train strength—his faux pas at the cabinet meeting notwithstanding. CEQ, Train wrote agency heads in one instance, "is concerned that, in a number of cases, there appears to have been failure to comply with the environmental statement requirement." CEQ, however, "is available to advise on questions that may arise."[31] Train remained firm yet understood many complaints. The Department of Transportation (DOT) noted that many "road decisions are made in the field and never get to Washington at all." Requiring statements for all such projects would "flood us with paper." Train responded that he understood and had "no desire to promote a huge paper mill." CEQ would do what it could to help but, nevertheless, such problems "do not remove from [DOT] the requirements of our statute."[32] Many agencies tried to transfer the responsibility for writing the statement to CEQ itself. Train refused but agreed to "a crash program to step up hiring and training in this field."[33]

To facilitate a smooth EIS process, Train instituted a system of CEQ "examiners" for each program area. In essence, this designated a given CEQ staffer to work with a specific group of agencies, allowing the staffer a greater expertise and all involved a better working relationship. Perhaps looking for partisan advantage, many critics complained that CEQ frustrated citizen participation. Train had a strong answer to such critics, including in the CEQ guidelines requirements for public hearings well in advance of any final action on the project. The EIS process, he ensured, was not just a bureaucratic formality.[34]

Train was not alone in strengthening NEPA. As all involved began to realize that the EIS was the unexpected cornerstone of the law, environmental organizations brought suits for failure to comply. Although they had not done so with the Amchitka nuclear tests, the courts frequently agreed and, in the process, expanded NEPA's scope. Such was the case when the Natural Resources Defense Council sued the U.S. Agency for International Development. A decision for the plaintiff meant, in essence, that international actions were not exempt. Train recognized that the expansion of NEPA's authority could make the process overly bureaucratic and burdensome. "I think many organizations, many businesses, many government units, will look back at [the EIS] with considerable concern because it does slow down the decision-making process," he told journalist Edwin Newman of NBC News. For the most part, however, Train hailed the new system, appreciating that it opened decisions to the scrutiny of private citizens and organizations. "Hopefully this will make for better planning, planning that involves the public, so that when a decision is made, the public will be fully aware of what the alternatives are and will in a sense share responsibility for the decision," he concluded.[35]

No agency could escape the importance of Train and CEQ. Not only did Nixon's executive order empower CEQ in the NEPA process, it also mandated that all federal agencies and departments comply fully with pollution laws by the end of 1972. Train had briefed Nixon on the issue two days before the order and convinced him that compliance was exceedingly poor. The task, Train knew, was monumental. The problem involved all federal buildings, public works, equipment, motor vehicles, aircraft, and ships. At Train's suggestion, Nixon agreed to spend $359 million to cover the requisite costs. The Department of Defense stood as the greatest culprit, Train explained. Over $3.1 million alone was necessary just to improve sewage that the U.S. Military Academy discharged into the Hudson River. Also at Train's suggestion, Nixon designated CEQ as the lead agency to monitor and ensure compliance. Once again Train looked over the shoulder of his fellow agency heads. Once again he found himself in the middle of dispute after dispute. Train, it appeared, was not about to leave the public eye.[36]

The spotlight was never greater than on the first Earth Day. While in recent years the day has been only a shadow of the original, the celebration on April 22, 1970, was a seminal event in American history, a fitting culmination to the growth of American environmentalism. Although the idea of an Earth Day originated with a young Stanford graduate student, Denis Hayes, and youthful activism drove the planning, the result was one of the largest demonstrations in American history. The concept caught fire among all strata of American society, uniting disparate groups often divided over the Vietnam War or the other momentous events of the time. Over a hundred thousand people took part in New York; an equal number participated on the mall in Washington. Congress adjourned for the day and television provided saturation coverage.[37]

The White House approached the day with apprehension, however. Some, such as Ehrlichman and Vice President Spiro Agnew, advised caution. The young organizers were often opponents of the war and harsh critics of administration policy. The event might devolve into a riot or violence, in which case the White House should disassociate itself. Others, however, such as Whitaker and Hickel, thought it prudent to endorse the day beforehand. They noted that Hayes had enlisted bipartisan congressional support and that ignoring it might undercut Nixon's environmental offensive. The solution, the White House finally agreed, was a policy of compromise. While Nixon would not take part directly, other administration figures should, in the words of Whitaker, "hit Carson, Cavett and the think-type programs." They should span out across the nation giving speeches and lauding the administration's environmental record.[38]

New CEQ chairman Train was obviously a man in demand, meeting Hayes and others as they made their plans. Train, not surprisingly, argued for a strong White House endorsement. Fielding requests from over a dozen universities to speak, including Duke and Cornell, he settled upon Harvard's Graduate School of Business Administration. Perhaps as a defense against possible radicalism, he stressed a policy of moderation and criticized environmental demagogues. "I am satisfied from discussions with reputable scientists that the available data simply do not support apocalyptic conclusions," Train argued. The need for action was urgent but the cause was not hopeless.[39] That night Train participated in a National Education Network program. Two days later he reiterated his message at the University of California at Davis. In the end, Earth Day spawned no violence, although partially clad students dancing and chanting just below the stage did interrupt Train's California speech. Later that week, Train joined White House staff in a symbolic cleanup of the Potomac River. His efforts did not dissuade critics, however, who noted Nixon's Earth Day absence and silence. This lack of credit frustrated Train but no more so than Nixon's failure to deflect it. "Nixon was unable or unwilling to do the simplest kind of overt demonstration of his commitment," Train recalled. "The waste treatment plant in Chicago was the only time I remember."[40]

At least in regard to Earth Day, Train might have forgiven Nixon. The National Security Council warned the President that communist supply lines through Vietnam's neighbor Cambodia threatened the war effort, which understandably dominated Nixon's attention. Nixon decided to invade that ostensibly neutral country, which caused massive domestic protests that dwarfed criticism of the administration's environmental record. Back at the Department of Interior, Hickel leaked his own critique of Nixon's actions. An infuriated Nixon decided to fire Hickel, ostracizing him until his official termination just after the November election. Nixon immediately began considering potential replacements, including several with no natural resources experience. This did not worry him. "Train could handle the substance," he assumed.[41] For his part, Train felt no joy in Hickel's political demise and surprisingly held no grudges. Hickel, however, did not always make it easy. Despite claims that he still wanted Train in his department, he had privately approved of Train's move to CEQ. When a reporter at that time asked if he were looking for a replacement with an equally strong environmental background, Hickel had answered half seriously, "Why should I look for a great conservationist when there's a great conservationist as Secretary of Interior?"[42] Now that it was Hickel's turn to depart, Train still sent a kind letter. "I want you to know how much I have enjoyed these past months in the Department of Interior and our work together," he wrote. Train tried to console his former boss, even forwarding him a copy of a newspaper article that essentially supported his position on Vietnam, which Hickel greatly appreciated. In later years, Train visited Hickel in his Alaska home. The two were never close friends but neither was Train one to remain angry for long.[43]

In any event, dwelling on the past was not an option when the present continued to demand so much. By the end of the year, Nixon settled on former Maryland congressman and Republican National Committee chairman Rogers Morton as Hickel's replacement, a man Train welcomed for his affable personality and decent environmental credentials. In the meantime, as the Hickel drama unfolded and the environmental bureaucracy settled, Train and his colleagues turned their attention to the myriad of environmental issues challenging the nation. NEPA mandated that CEQ issue annual reports, which consumed a great deal of Train's time. The first report in August 1970 proved particularly difficult. Expectations ran high with all the publicity. Train and his colleagues, however, faced considerable constraints. The law set a deadline of July 1, yet the initial efforts to establish CEQ proved time consuming as well. When it became apparent that CEQ would miss the deadline, Train wrote Whitaker to apologize. "We are working night and day now," he noted.[44] Train wanted the report to win praise but remained aware both of fiscal constraints and of the fact that Whitaker's task force had already outlined much of the administration's first-year agenda. Following the rebuke over his SST remarks, he did not want to rock the boat. "We are fully aware of the current fiscal situation and the hazards of committing the Administration to programs that might be difficult to fund," he wrote. "We have cast these recommendations as general program directions without indicating specific deadlines and program costs." Train did not have time to consult adequately with BOB and, therefore, included no major legislative proposals. "We feel that the program directions in this report are the least we can do to comply with the provisions of the Act."[45]

With the CEQ report essentially reiterating what Whitaker's task force had already proposed, criticism was inevitable.[46] Some newspapers praised the report. The *New York Daily News,* for example, called it "sort of a traveler's guide to a better, more livable country." Others, however, expected more from Train. While noting that CEQ had less than three months to prepare, the *New York Times* dryly reported, "The document gave an exhaustive analysis of the ecological problems facing the country but offered few new proposals for solving them."[47] When the administration officially presented the report to Congress, Train testified and noted the White House proposals already on the table. A certain defensiveness was evident in comments to journalists. "Very few people seem to appreciate the complexities with which we must deal," Train wrote the *Saturday Review.*[48]

With more time to prepare, CEQ presented a more detailed report the following year. CEQ also assumed in 1971 the responsibility for writing Nixon's annual environmental message to Congress, presented after each year's State of the Union address. Published as the "President's Environmental Program," the messages afforded CEQ more opportunities to define publicly the administration's agenda. Now that Whitaker's task force no longer composed the message, CEQ's annual reports were never again redundant. Presenting CEQ's draft of the 1971 message, Train was blunt with Nixon: "You now have a unique opportunity to build an environmental program that in the space of the next several years could turn around the way we handle our environmental problems." The press uniformly praised the report, although the *New York Times* declared Nixon's reaction to it "timid."[49] Submitting the annual reports and messages often provided Train a chance to meet with Nixon. In one instance, Nixon commented to Train that he had just seen the musical *Hair.* "That has a lot of nudity in it, doesn't it?" Train asked. "What's wrong with a little nudity?" Nixon shot back. Nixon frequently sat in a chair near the fireplace, Train recalled, the fire always burning regardless of the temperature outside. To Train these meetings were "stroke sessions," in which Nixon tried to impress upon him the White House's commitment. They rarely were substantive but, to the CEQ chairman, important nevertheless.[50]

Whether reflected in the various reports, the opinions of others, or the occasional meeting with Nixon, Train and his colleagues developed an impressive record of substantive accomplishments in CEQ's early months, further indication of Train's clout. Most impressive was CEQ's influence in halting the Cross Florida Barge Canal (CFBC), then six years into construction. A canal cutting across central Florida promised significant economic benefits and, accordingly, won the unanimous support of Florida's congressional delegation. As Nixon undoubtedly knew, a large percentage of the electorate in this key political state supported continued construction. The momentum for the project did not daunt Train, however; his tenure on the National Water Commission lent him a level of expertise. CEQ first recommended a suspension of work pending a study of relevant environmental issues. Soon it reported that these studies not only confirmed the worst-case scenario but also challenged the potential economic benefit. "This project could seriously affect the environment in Florida by degrading water quality," Train explained. "The project is marginal from an economic point of view and hence very undesirable in the face of the potential and actual environmental problems it presents."[51] Reassuring critics and politely addressing the issue with congressional supporters, he argued against any potential realignment as a compromise. Backing his argument with specifics, Train convinced Ehrlichman, who in turn convinced Nixon. In January 1971 Nixon ordered a halt to construction. Referring to the CFBC as a "past mistake" and the area a "natural treasure," Nixon declared, "I have accepted [CEQ's] advice."[52]

Train had more advice for the problem of trash and other solid wastes. After Whitaker's task force convinced Nixon to advocate more funds for solid waste management and recycling, Nixon instructed CEQ to recommend ways to deal with abandoned automobiles. When word spread of Train's task, the National Center for Solid Waste Disposal pressured him to move slowly, claiming that the industry needed time to prepare. Train, however, refused to budge: "I said that I could give no such assurances and that, if anyone came up with any good ideas, I assumed that we would proceed." Within months, Train and his colleagues had their "good idea"—a federal loan program to help businesses purchase modern metal scrap shredders. The money, drawn from an increase in the auto excise tax, would also fund a program to improve scrap recycling.[53] If this were not enough, CEQ proposed a ten-cent deposit on all bottles and cans to encourage beverage-container recycling. This latter proposal dragged Train into negotiations with NIPCC and William Austin, CEO of Coca Cola. Although in the end Nixon rejected both proposals, Train's advocacy was still influential. When Congress passed the Resource Conservation and Recovery Act of 1970 late in the year, a bill that Muskie sponsored and that included significantly more funds for solid waste management than the White House thought appropriate, Nixon signed the legislation rather than vetoing it. Agreeing with Train, he knew that the administration was vulnerable on the issue. As an amusing bit of irony, Train had his own waste disposal problems as Congress debated the legislation. Unable to get Washington's sanitation department to pick up the trash from his home, he employed his connections once again. "I am becoming desperate about the quality of my personal environment," he wrote Mayor Walter Washington.[54]

In another form of recycling, Train played a key role in the Property Review Board (PRB), an innovative body whose purpose was to find uses for surplus federal property. The PRB was to identify all underutilized General Services Administration property and make recommendations for disposition. Much of this property would become parkland, often in urban areas virtually devoid of nature. In this capacity, Train developed a relationship with Donald Rumsfeld, later Secretary of Defense under two administrations but then PRB chair. In what must have been a satisfying moment, Train represented the government when it turned over the place of his birth, a military base, for a state park.[55]

The two most important substantive issues facing CEQ were, of course, air and water pollution, the former difficult to miss in the summer of 1970, with record smog blanketing a broad stretch from Washington to Boston. Whitaker's task force had proposed national ambient air quality standards, obtained by having the states set specific emissions standards for stationary sources and by raising the emissions standards for 1973–75 model automobiles. Nixon liked the idea and it passed the House of Representatives only to bog down in the Senate. There Muskie proposed a competing bill, similar but with much stronger automobile standards. It mandated a 90 percent reduction in hydrocarbons and carbon monoxide by 1975, and a 90 percent reduction in nitrogen oxide by 1976. The automobile industry claimed such standards were impossible to meet, while environmentalists argued that they would spawn innovation.[56]

As Congress debated, Train argued for tough provisions, once again appealing to Nixon's political instincts. "It gives [Nixon] visibility on the air pollution issue," Train wrote Ehrlichman. Train and his colleagues also advocated a tax on leaded gasoline. "I refuse to believe that breathing lead particulates is not damn bad for me," he concluded. After Train testified in favor of such a tax before his old colleagues at the Ways and Means Committee, Nixon agreed and the tax passed.[57] At the same time, Train pushed for the development of a pollution-free automobile, convinced that the major companies were dragging their feet in this regard. He lunched with a leading advocate, independent automaker William Lear, and lobbied for additional federal research funds.[58]

As the end of the year approached, Congress passed the Muskie bill and Train worried that Nixon might veto it. Reminding Nixon that the bill passed overwhelmingly, Train tactfully appealed to Nixon's ego. "The basic features of the Senate bill were derived from the Clean Air Act amendments you proposed to the Congress in your Message on the Environment," Train wrote.[59] Nixon, in fact, decided against a veto. Instead, perhaps taking a clue from Train, he sought to take full credit for the bill. On the last day of the year, Nixon signed the monumental Clean Air Act Amendments of 1970. In a press conference after the ceremony, Ruckelshaus boldly declared that the White House welcomed the emissions standards and had not worked against them. Notably absent from the ceremony was Muskie, whom Nixon had conspicuously not invited. An outraged Muskie protested, with the resulting press frenzy ensnaring Train. A reporter asked Train if it were unusual not to invite one of the key architects of a bill. Train had played no role in the snub and thought it "petty." He had to answer in the affirmative, however, thus fanning the flames of outrage. The White House fought back, dispatching Train to appear on *Meet the Press* to tout the administration's record the best he could.[60]

The debate over water pollution legislation was even more contentious. Whitaker's task force proposed $6 billion in federal funds for the construction of waste treatment centers and precise federal affluent standards, the latter a significant break from the ambient standards employed previously. The bill extended jurisdiction to "all navigable waters," not simply those clearly interstate. As if to challenge the administration again, however, Muskie proposed an alternative. The Muskie bill maintained many of the same provisions but more than doubled the federal funds. Nixon had just signed legislation that mandated absolute company liability for oil spills, an action Train supported. Now, however, Nixon insisted upon the lower appropriations in the new bill. Debate over the level of funding soon stalled the bill in Congress, frustrating environmentalists who sought quick action.[61]

Train recognized an opportunity. Following a Supreme Court case that held "refuse" to include liquid wastes, he suggested using the seventy-year-old Refuse

Act of 1899 to battle pollution. Never intended in this way, the law established a Corps of Engineers (COE) permit program to prevent the discharge of materials from blocking the passage of ships. Train convened a meeting in CEQ's reference library and conference room, loaded with books that he had selected. Train envisioned COE issuing permits according to Federal Water Quality Administration standards. With Ruckelshaus present at the meeting, Train stressed that EPA should play a role "once fully up and running." "We should not foreclose consideration of a program which would ultimately place the permit authority in EPA," he argued.[62] Whitaker liked the idea because it implied executive action in the face of congressional intransigence. It might also encourage Congress to act. In any event, Train assumed the program was temporary. "It wasn't a very satisfactory substitution for legislation because, from an administrative standpoint, it was something of a nightmare." Only a week before Nixon signed the Clean Air Act Amendments, Train announced the program before the White House press. The following day newspapers carried the story on their front page, the *Washington Post* including a picture of a thoughtful Train, hand to his chin, answering questions.[63]

Other avenues for action existed, Train argued. When the White House recommended that CEQ study the problem of ocean dumping, Train convened a task force with members across a number of federal agencies and departments. He knew that dumping dredge spoils, municipal sewage, industrial debris, and various chemical and radioactive materials posed a potential time bomb. "We are strongly of the opinion that ocean dumping if not checked can lead to serious environmental consequences," he wrote Congressman John Dingell.[64] When Train's task force recommended a ban on the dumping of certain dangerous materials and an EPA permit system for others, Nixon welcomed the report and promised specific legislation early the following year. Train stressed that the United States should work with other countries. In presenting his report, he noted the issue as an "important opportunity for international leadership."[65]

Train did not have to look far to recognize the need for international action. Reports from Lake Erie spoke of a "mat of algae two feet thick" during the summer months. With a minimal flow of water and a large industrial base along their shores, the perilous condition of the Great Lakes demanded a coordinated response from the United States and Canada. Phosphates, an integral ingredient in detergents, were a major cause of eutrophication but a point of contention between the two nations. Using the chemical substitute nitrilotriacetic acid, or NTA, the Canadians pushed for a complete phosphate ban, a move environmentalists applauded but the Nixon administration rejected. As environmentalists pointed to the power of the detergent lobby, the White House claimed NTA as a potential carcinogen. Complicating matters was the simple fact that most of the industrial base stood on the American side.[66] In addition, the report of the International Joint Commission (IJC) added pressure. Established under the 1909 Boundary Waters Treaty, the IJC repeated the Canadians' call for a complete phosphate ban.

Train wasted no time. He first met with the three largest detergent producers—Procter and Gamble, Colgate-Palmolive, and Lever Brothers—and pushed for voluntary phosphate reductions. When only one company agreed to make a significant cut, Train reported the failure to Nixon. "We do not believe these proposed actions are significant enough to sell as a government-industry voluntary agreement," he wrote, countering a report by the Department of Commerce. The solution was an "incentive tax" phased in over five years. Such a tax "does not put the Government in the difficult position of prescribing phosphate levels."[67] Nixon, however, balked at the idea. Always reticent to raise taxes, he left Train in a difficult position.

Nixon had already appointed Train the chair of a task force to address the problems of Lake Erie, and Train took the opportunity, first, to expand his study to all of the Great Lakes and, second, to approach the Canadians to join him. Employing the diplomacy for which he was well known, Train got along well with his Canadian counterpart, Secretary of State for External Affairs Mitchell Sharp, who was several years older but shared his sense of humor. Their personal relationship was important because the hurdles facing their U.S.-Canada Joint Working Group (JWG) included the Canadians' insistence that any agreement allow both countries to discharge the same amount of pollutants. "We emphasized at the meeting that the policy of both nations should be on positive pollution abatement rather than on dividing up pollution," Train wrote Nixon, both aware that an equal distribution placed more of an onus on the Americans than the Canadians.[68]

Almost a year passed before a tentative agreement was achieved. Finally, in early June 1971 the JWG issued its report. The report called for a joint water quality standard, an 80 percent reduction in phosphates by 1975. To achieve this goal both pledged to build a large number of sewage treatment plants. The agreement also called for the total elimination of mercury and other toxic substances and better control of radioactive wastes and pesticides. Train and his colleagues called the agreement "an historic first step," recognizing that if both nations officially signed the pact, it required significant costs. A formal signing, they promised, would follow in several months.[69]

Train's plea for international cooperation in the fight against pollution found a receptive audience in the commander-in-chief. Nixon shared Train's love of travel and international relations and prided himself on his diplomacy. The environment might work to his advantage politically, he assumed, but it might also serve as a bridge to other nations in the bipolar Cold War world. The result, for Train, was a return to his international conservation roots. Once again he was a world traveler, a man whose commitment to the environment—and, now, to his country—demanded time abroad. Ruckelshaus had graciously ceded international environmentalism to Train, who not surprisingly welcomed the responsibility.[70]

Train benefited from Nixon's own brand of diplomacy as well. Nixon eschewed the bureaucracy and favored personal diplomacy. He did not rely on the Department of State as much as previous presidents, depending instead on individual emissaries and the smaller National Security Council (NSC).[71] Nixon recognized that Train's background and interpersonal skills made him a perfect diplomat. "The White House very much looked to me and CEQ, not the State Department, to handle international aspects of the environmental program," Train recalled. "I think the same was true in other areas like [Henry] Kissinger at NSC."[72] Ever astute in the ways of Washington, Train recognized the potential for friction with his colleagues at the State Department and worked to diffuse problems before they arose. He requested that State attach a foreign service officer to CEQ as a liaison. He also kept the department informed of his initiatives and deferred to it when necessary. In time he developed a strong relationship with Christian Herter, Jr., a top departmental employee. Train was obviously a diplomat in more than one way.[73]

Train's environmental diplomacy afforded him additional opportunities to visit with Nixon. Occasionally he would meet Nixon in the Oval Office before departing. Train welcomed such sessions because they allowed him to tell the country of his destination that he had just met with the President, implying of course that he spoke with the highest authority. In one instance, Train apologized to Nixon for invoking his name to influence negotiations. "No," Nixon replied, "that's exactly what you should do."[74]

In the summer of 1970, the United States was not the only country suffering through hot hazy days of smog. Rapidly industrializing Japan shared the same fate and, with almost no environmental bureaucracy, was an obvious nation of concern. Train and CEQ suggested that Nixon contact Prime Minister Eisaku Sato to begin cooperative efforts. Nixon agreed and wrote Sato. Not surprisingly, Train soon headed a delegation to Tokyo. First, however, came a meeting with Japanese Ambassador Nobuhiko Ushiba. When Ushiba joked about the publicity surrounding Train's orangutan photograph, Train good-naturedly promised him a signed copy.[75] Once in Tokyo, Train met with Sato and Sadanori Yamanaka, the country's environmental coordinator newly appointed for the American's visit. The Japanese publicized the meeting with skits of mythological figures poisoned by pollution. Discussions revolved around the need to balance economic growth and environmental protection. One minister suggested that the United States might use environmental restrictions to bar the importation of foreign goods while another wondered aloud whether the "real reason" for the visit was to drive the price for Japanese products higher. According to the Japanese press, when Train noted the pollution in Tokyo Bay, "the Prime Minister was put in an awkward position and the only thing he could do was to force a smile."[76]

For the most part, however, the meeting was both friendly and productive. The Japanese agreed to join the United States in fighting ocean dumping and to continue bilateral discussions. Sato soon met with Nixon, including environmental protection as part of their agenda. In time, Japan followed America's lead and created its own Environmental Protection Agency, imitation the surest form of flattery. Train returned to Washington proud of his accomplishments and with more memories for his scrapbook. Aileen, who accompanied her husband, joked about watching him give an interview on Japanese television. Not understanding the Japanese voice-over, she was unsuccessful in reading his lips. The couple indulged in some sightseeing at Akan National Park. There they enjoyed a tremendous view reminiscent of a traditional Japanese painting, but only after passing piles of trash to gain their vantage point. Train characteristically thanked State Department employees who assisted with the logistics. "I really do appreciate all the time and effort you put into handling our complex negotiations for rooms, gin, statues, ice, massages, etc., during the trip," he joked.[77]

Once back in Washington, Train received a call from the Spanish ambassador. King Juan Carlos planned an official state visit and wanted to schedule a meeting to discuss environmental affairs. Train naturally accepted and met Carlos at the Blair House, the government's official guest residence. Although the meeting lacked much substance, the king invited Train to Spain, a visit he enjoyed the following year. There Train met Francisco Franco. Franco was old and weak, battling Parkinson's disease and holding his arm to keep it from trembling. Again the meeting lacked much substance but, for Train, was memorable nevertheless. Given the Spanish Civil War and his long rule, Franco was "somewhat of a legendary figure to my generation." Train may have understated the importance of his visit because in the years that followed, Spain too developed its own environmental bureaucracy.[78]

Train dutifully reported to the White House the results of his trip, Nixon sharing Train's interest in the historic figure. The President, Train recalled, always admired strong leaders, if not their policies and politics. In one meeting Nixon stressed the need for forceful authority by reading a passage aloud from Albert Speer's *Inside the Third Reich*, an account of the armaments industry during Nazi Germany. It was an odd and unfortunate reference, Train thought, later recalling the quote as the Watergate scandal unfolded.[79]

Train's international reputation continued to grow, overseas and at home. As he departed for Spain, France also invited him for an official visit, only to have his full schedule prevent it. Nevertheless, he met with both the French ambassador and the minister of agriculture, encouraging them to stress environmental protection. They could expect industry to resist, Train explained, undoubtedly reflecting upon his own experiences. In time, France joined her neighbors in strengthening her environmental laws.[80] At home, Nixon appointed Train to head the American delegation to NATO's Committee on the Challenges of Modern Society, replacing Daniel Patrick Moynihan, who had served since Nixon first proposed the committee the year before. Nixon also proposed in his 1971 environmental message the creation of the World Heritage Trust, for which Train had lobbied since his days at the Conservation Foundation. Train's position brought him and Aileen to the occasional White House dinner and social function. In one instance Aileen and Nixon discussed what it took to govern a city. Several months later she greeted him in a receiving line only to have him recall the conversation. While the comments demonstrated Nixon's intelligence, they were also evidence that Train and his cause had arrived.[81]

As his fifty-first birthday approached in 1971, Russell Train appeared to be at the pinnacle of his career, enjoying his influence on an important issue that galvanized the world. He had the President's backing and strong public support. Republicans praised him. He contributed money to the Grand Old Party and advised its leadership, even as he maintained his friendship with Democrats.[82] Most environmentalists joined in the praise. Adding to his long record of advocacy and accomplishment, he publicly fought the Internal Revenue Service

(IRS) when it attempted to revoke the tax-exempt status of organizations that filed pollution lawsuits, an action of obvious importance to the full spectrum of environmental community. He wrote the IRS and, in the end, prevailed. His alma mater awarded him an honorary degree, where he joined such notables as Coretta Scott King and singer Bob Dylan. Even his neighbors praised him as he successfully fought the city government to maintain trees on nearby land.[83] Known both around the block and around the world, he was a success as both a Republican and an environmentalist. He remained lionized, working hard to keep the lions at bay.

6

The Tide Turns

"Throughout my life," Train recalled, "I've been a tremendously lucky fellow."[1] In the summer of 1971, as he entered his sixth decade, little had happened to suggest otherwise. Health, wealth, and most recently, notoriety surrounded a successful career dedicated to the common good. Train had worked hard but fate had placed few major obstacles in his path. His good fortune, however, was about to change.

It all began with Nixon. Over the years Nixon had competed with two leading Republicans to carry the banner of the party, New York governor Nelson Rockefeller and Arizona Senator Barry Goldwater. In many respects, the former represented the traditional, northeastern wing of Republicanism, which favored active, if not better-managed, government. The latter represented a growing number in the West and South, more ideologically driven and hostile to federal power. Nixon remained in the middle. Caring little for domestic policy, his attention focused on the international stage, his administration aimed to unite the party. He delegated significant domestic authority to subordinates with only the broadest mandate for moderation.[2]

The result was, arguably, a rather convoluted record. On the one hand, as his first term passed its halfway point, Nixon had abandoned the harsh rhetoric and abrasive tactics for which he was famous. He had in many respects fulfilled his election promise to "bring the American people together."[3] His cabinet was largely moderate and won praise, Hickel the lone exception, and his first budget hardly attacked the Great Society.[4] While he supported welfare reform that instituted work requirements, he also increased federal expenditures. He indexed Social Security payments to the rate of inflation, expanded consumer and employee protections, supported lowering the voting age to eighteen and signed the Equal Rights Amendment. According to his speechwriter, William Safire, Nixon's "heart was on the right while his head was with FDR, slightly left of center."[5] On the other hand, however, Nixon coupled this surprising spate of liberalism with actions pleasing to conservatives. He nominated several Supreme Court justices hostile to civil rights. He advocated "law and order," a new "war" on illegal drugs, a crusade against obscenity, and opposition to both abortion and forced busing. His government reorganization stemmed from a genuine desire to decrease bureaucracy and he championed "revenue sharing," allowing states more autonomy in the expenditure of funds. In total, many of the administration's actions appeared a hodgepodge of both liberalism and conservatism, a convoluted policy indeed.[6]

The result may have been contradictory overall but was just the opportunity that allowed Train to flourish. Appealing to Nixon's partisan instincts even as he encouraged the moderate wing of the party—the traditionalists more open to the bipartisanship he valued—Train helped craft the record of accomplishment that most environmentalists applauded. Had Nixon assumed the Presidency with a firm domestic agenda or the commitment to conservatism that pundits predicted, Train's story would have been different. Train was, in short, fortunate.

By the middle of 1971, however, Nixon, increasingly sensitive to criticism, had begun to have doubts. *Washington Post* columnist David Broder wrote that Nixon "seemed more unpredictable, more mysterious, more inconsistent" three years into his term than at the outset. Voters were "alienated," Broder concluded, with "trust and confidence" the big issues in the pending presidential election.[7] Ben Wattenberg's 1970 text, *The Real Majority*, argued that Republicans could transform American politics by appealing to the "unyoung, unpoor, and unblack."[8] As his advisors cited the book, Nixon warmed to his famous "Southern Strategy." He could win with white southerners, northern Catholics, right-wing labor leaders, and other socially conservative Democrats alienated by the 1960s counterculture. He could drive a wedge between the "silent majority" and upper-class liberals who dominated the Democratic establishment.[9] Nixon had touted such a policy for months and its roots were in many of his conservative first-term accomplishments. Now with the 1972 election approaching, however, he might more forcibly tack to the right, perhaps alienating many moderate Republicans but winning more votes in the aggregate.

Challenging bureaucracy and demonizing regulations did not bode well for the environment—or Train. "The environment is not a good political issue," Nixon told Haldeman soon after the midterm elections. "We are doing too much," he concluded, "catering to the left in all of this." The administration could never win the environmental vote. "You can't out-Muskie Muskie."[10] While largely masking his shifting attitudes in public, Nixon was blunt in private. "I have no sympathy for environmentalists who are demanding equal time on the air for every reply to every issue," he declared. "Some people want to go back in time when men lived primitively." By the summer of 1971, Nixon increasingly referred to environmentalists as "dippy" and the "wacko fringe."[11] Industrial lobbies and their champion at the White House, Maurice Stans, encouraged such attitudes. Business, Stans claimed, had become a "whipping boy for the environment." Environmentalists had "hostile public attitudes about business." Nixon agreed. The environment, he now concluded, "was no sacred cow."[12]

Ever the astute politician, Nixon did not want to alienate environmentalists unnecessarily. The best tactic was, in the words of one White House aide, to "maintain to the extent possible the position of continuing concern for the environment, but stay away from specific issues."[13] The administration needed cover in its retreat; it should take what actions were possible without hampering the economy or angering conservatives while continuing to pledge fidelity to the cause. Nixon, therefore, still needed Train, who still had an opportunity to advance his agenda. In the years that followed, Train continued to accomplish much on behalf of the environment, even as his defeats now matched his victories. To much of the public Nixon's environmental policy appeared as convoluted as the rest of his domestic policies. Indeed, even as Nixon more vigorously preached the virtues of deregulation, he instituted wage and price controls. Inside the White House, however, no doubt existed over the new direction. An increasingly frustrated Train had to balance his desire to remain in the administration, to remain part of the team, with his desire for the most forceful environmental policy possible. Environmentalists demanded the latter and were chagrined when Train succumbed to the former. The beleaguered CEQ chair worked as diligently as ever, the task of being both a Republican and an environmentalist growing more difficult by the day. His fortune had begun to turn.

Train knew where he stood. In one early 1973 meeting at the White House, Nixon expressed exasperation with the "hand-wringing on environmental problems." Americans, he claimed, "don't really get upset about air and water pollution." When Train explained the importance of proper land management, Nixon blurted out, "Christ, I wouldn't want to live on a farm." When Train pressed his case, Nixon recalled that he played golf with friend Bebe Rebozo in Key Biscayne. The pair found a row of mangroves blocking their view. "We wanted to cut just a few vistas through the mangroves but were not allowed to," Nixon stated. "Yet these mangroves aren't good for anything except gooney birds." As Train sat listening, amazed at the lack of sophistication in Nixon's perspective, White House valet Manuel Sanchez entered the room carrying coffee. "Isn't that right, Manuel?" Nixon asked. Sanchez's reply showed a certain gumption lacking in many of Nixon's closest advisors. "No, Mr. President," came the answer. "You know that when I have my day off down there I go fishing. I know that the fish I catch need those mangroves to grow up in. If you cut them down, there won't be any more fish." Later, after leaving the room, Train cornered Sanchez. "Maybe the President will listen to you," Train stated. "He doesn't really believe me on this." Recalling the incident, Train mused, "It is hard to know how to make [Nixon's] view of environmental matters more knowledgeable and sophisticated."[14]

Nixon may have concluded that air and water pollution did not resonate with the public, but at least one—water pollution—still dominated Train's agenda. With the passage of the Clean Air Act Amendments the previous year, the focus on the nation's air had turned from the formation of policy to the implementation of law. It was not CEQ taking the lead, therefore, but EPA. It was not Train that drew the attention but his friend Ruckelshaus. In his remaining years at CEQ, Train and his colleagues offered only one new air pollution proposal—a tax on sulfur oxides, which the electric power, smelting, and refining industries emitted. While Nixon initially accepted the idea, it predictably died in Congress, the victim of powerful industrial lobbies and a degree of administration lethargy. For the most part Train remained quiet in regard to air pollution; the cause was left to others.[15]

Water pollution was another story. With the new Congress in 1971 came a renewed battle over federal funding for waste treatment centers; the debate between the administration's bill and Muskie's competing, more expensive proposal was left unresolved from the previous session. A new degree of urgency existed. The Refuse Act permit program, which Train had touted the previous year, had proved a debacle. While Train had predicted that EPA would soon issue the permits for discharges, the Corps of Engineers (COE) remained in control, issuing permits at a rate that undermined legitimate environmental protection.[16] Swamped with over twenty thousand applications in only several months, COE granted permits with only superficial study, a "license to pollute" in the words of one environmentalist. Train refused to join the environmental community, however, citing several high-visibility prosecutions under the law, including U.S. Steel and Mobil Oil, as a defense of the program. He no longer strongly pushed for EPA jurisdiction and did not complain when the administration announced that the program would only apply to interstate waters, not all navigable waters as he had initially promised. Environmentalists responded by turning to the courts, arguing that each permit required an environmental impact statement (EIS) under NEPA. Once again Train took a position in defense of the failed program, arguing for an exemption. The Refuse Act program deserved "our strongest support," he claimed, "but implementation should not be rigid or inflexible."[17] It was an ironic twist—a champion of NEPA now, at least in this regard, attempting to limit its impact. Train's argument did not carry the day, however; the U.S. District Court for the District of Columbia ruled in the environmentalists' favor. The result, both Train and the environmentalists knew, was the death knell for the permit program.[18]

The pressure was now on Congress. When Muskie reintroduced his $25 billion program to build waste treatment centers, the federal government covering half, the Nixon administration responded by increasing the funding it proposed to $12 billion, also half from Washington. Unlike the Muskie proposal, the administration required only the "best practical technology," not the "best available technology," wording that allowed companies to argue that certain technologies were financially impractical. Both bills kept at the expense of states strong federal jurisdiction over standards and a permit program to achieve them. Both were dramatic improvements over existing law but, arguably, failed to address the full extent of the problem. The National League of Cities and the U.S. Conference of Mayors claimed that greater sums were necessary, while independent studies placed the total costs between $33 and $35 billion.[19]

Years later Train had little to say about his role in the debate. "Personally," he recalled, "I don't have any recollection of my opinion," an amazing admission from an environmentalist in one of the nation's most important environmental debates.[20] In reality, Train, fully aware that Nixon had just declared environmentalists were "going crazy" over the issue, did not want to alienate himself from the White House. Angering those who put so much faith in him, Train testified with Paul McCracken, former chairman of the Council of Economic Advisors, in favor of Nixon's lower costs and less-stringent provisions. Industries, Train argued, might shut down or move to foreign countries if Muskie's bill passed. Inflation might result and, worse in his view, "public acceptance for environmental programs" might wane. Moving too far, too fast meant "water quality in many cases may suffer."[21]

Such a scenario was doubtful, but the conservative in Train meant what he said. He genuinely worried that increasing costs might spawn a public backlash and, even as he fought the Muskie provisions, sought to assure the public that it could afford environmental protection. Just as he protested the $25 billion that Muskie proposed for waste treatment, he also, ironically, claimed that the nation could afford environmental costs that could surpass $100 billion by 1975. "This amounts," he stated, "to less than one percent of the gross national product over [the same period]."[22] Taken in total, Train's advocacy appeared almost as convoluted as the administration's overall domestic policy. It also may have played into his opponents' hands. In a sense, Train's insistence that the nation could afford environmental protection in general undermined his and the administration's specific objections to the Muskie water pollution bill. In the end, Congress passed the Water Pollution Control Act Amendments of 1972, which kept most of the Muskie provisions and even increased the percentage of federal funds to $18 billion.

A swift veto followed. CEQ joined EPA and the departments of State and Interior recommended approval, but an irate Nixon was adamant. Unlike Ruckelshaus, who strongly lobbied against the veto, Train was surprisingly quiet. It did not take long for Congress to override the veto, a defeat for the administration and a victory for environmentalists. Train, his advocacy so strong in other areas, proved disappointingly weak in this one long struggle. It was not his finest hour.[23]

In other aspects of water pollution, however, Train was truer to form. When COE proposed building a dam on the Delaware River, the Tocks Island Dam, Train was resolute. In a struggle that rivaled the Cross Florida Barge Canal fight the previous year, the project cost several hundred million dollars and promised water and energy for four states. It also created a lake with over thirty miles of shoreline, a huge body of water sure to accumulate the agricultural runoff from the farms upstream. Train knew the result—eutrophication that would kill all aquatic life. COE's EIS relied on the states' promises to address the issue, assurances Train and his colleagues thought inadequate. Demanding more specifics, Train insisted that CEQ return the EIS pending stronger environmental safeguards. Months passed, negotiations commenced, but Train would not give an inch. "I told [COE] that I strongly recommend against letting of the bids until the assurances requested have been received," Train reported.[24] When proponents issued a revised EIS, CEQ returned it again; it still lacked specific assurances, Train believed. As the New York Times praised Train, COE ultimately surrendered. The final result was not a dam but a national scenic river.[25] Following upon this success, Train also proposed adding Florida's Oklawaha River to the National Wild and Scenic Rivers System. Although Congress balked, it was another proposal that environmentalists appreciated, an action more in line with what they expected from the CEQ chairman.

As the debate over the Tocks Island Dam raged, Train gained a rare opportunity to implement and enforce water pollution law, expanding beyond the realm of policy formation. The White House appointed him "federal coordinator" to clean up asbestos pollution in the waters around Duluth, Minnesota. The Reserve Mining Company dumped tons of taconite tailing daily into Lake Superior, spreading the known carcinogen. Aware that he risked resentment by treading in EPA's domain and reflecting again his deftness in the ways of Washington, Train wisely gained the support of the agency's regional administrator. He also coordinated his efforts with Duluth's city government, even cornering the mayor in Chicago's O'Hare Airport. He finally organized his own staff for the task, dispatching assistant William Muir to serve as his on-site representative.

Convinced that such organization was sufficient, Train demanded that COE send an "erdalator," a newly developed water purifier, to the city. He mandated the National Water Quality Laboratory study the issue and also ordered a review of alternative drinking water sources. Train's pressure on the mining company was immense, which environmentalists, still simmering from their dispute over funding for waste treatment centers, applauded. When a lawsuit the following year ordered an end to all taconite tailing discharges, Reserve Mining finally decided to shift its disposal to secure land sites, a victory for the Great Lakes and, in environmentalists' eyes, for Train.[26]

Train had high hopes for the Great Lakes, anticipating that the administration would accept the terms he had earlier helped negotiate with Canada in regard to phosphate pollution. He had long labored in the U.S.-Canada Joint Working Group to achieve a strong joint water quality standard, an 80 percent reduction in phosphates by 1975, and hoped that the official signing ceremony for the Great Lakes Water Quality Agreement would continue to smooth the ruffled feathers of environmentalists. When the ceremony finally came in 1972, however, the terms did not completely fulfill Train's promises. The final agreement spoke only of "water quality objectives," relegating specific requirements, at the administration's insistence, to the appendix. These requirements no longer promised an 80 percent phosphate reduction by 1975, but only 60 percent. If Train were disappointed, he chose not to show it, traveling with Ruckelshaus to Ottawa for the official signing ceremony. Later he declared efforts to improve the Great Lakes "one of the greatest success stories in American history." The Great Lakes Water Quality Agreement was doubtless a dramatic improvement in the fight against pollution, even if Train's praise overlooked the weakened provisions. Given the criticism he had recently endured from his own environmental allies, however, one can understand his efforts to make the most of the situation.[27]

Train never expected his job to be easy, with water pollution only one of his frustrations as the 1972 election approached. A greater disappointment loomed, a failure with, arguably, more significant environmental ramifications. As the administration wrangled over funding for waste treatment and levels of phosphate reduction, millions of acres of rural America continued to succumb to the bulldozer and concrete mixer. Development sprawled out across the land with virtually no controls and no assurances of environmental protection. Although the power to zone emanated from states' police powers, most states had delegated such authority to counties and cities. This meant that no adequate land use policies existed, no way to protect ecologically fragile areas that extended through several local jurisdictions.[28]

Problems were apparent from the start. When Washington Democratic Senator Henry Jackson introduced a strong bill mandating state-wide plans with grants for assistance and penalties for noncompliance, the White House might have endorsed it but chose to remain quiet. Worried over the costs and sanction provisions, it ordered Train and CEQ to develop an alternative. In this way the White House could remain above the fray, as conservatives in Congress were sure to object. "Jackson can't be against Russ working on it and holding hearings since Jackson 'created' Russ' council," Whitaker told Ehrlichman. "Let's punt using Train as the excuse."[29] Working with Boyd Gibbons and William Reilly, Train struggled to craft his own proposal. "The larger questions of land use," Train acknowledged in a letter to Senator Alan Cranston, "are among the most difficult of our environmental issues."[30] Not only were there fiscal concerns but questions of jurisdiction. George Romney, Secretary of the Department of Housing and Urban Development, and Rogers Morton, Secretary of the Department of Interior, both sought control of the program, meeting with Train in his office. Other departments complained as well. James Beggs, Under Secretary of the Department of Transportation, complained to Train that he thought his department had been left out of the debate. "As I recall," an obviously annoyed Train replied, "I arranged representation directly with you." CEQ, he continued, "does not intend to push any proposal down any agency's throat."[31]

"Land use is the nation's top environmental priority," Train increasingly argued. "Land use issues lie at the heart of many of the critical environmental decisions facing the nation, whether they be air quality implementation plans, decisions on where to locate large-scale energy facilities, policies for the use of our public land, how best to manage our national parks and forests, seasonal home developments in the mountains and along the coasts, historic preservation, strip mining, stream channelization and urban sprawl."[32] The issue was important enough to warrant a White House conference, he claimed. With questions of pollution shifting from Congress to the courts, land use represented "the most important environmental issue remaining substantially unaddressed as a matter of national policy."[33] The White House, however, was once again reticent. "I'm frankly suspicious," Whitaker wrote Ehrlichman. The conference would be "a built-in chance for the doomsday boys to say we are not doing enough." Ehrlichman agreed; the administration would "duck" Train's proposal.[34]

When CEQ finally presented its proposal, conservative concerns were obvious. The proposal called for $20 million annually for five years to aid states in the development and implementation of land use plans. Given the complete lack of preparation in almost all states, however, the $100 million total was woefully inadequate, a fact that even the administration's Office of Management and Budget acknowledged. More troublesome, however, was the mandate to cover only areas of "critical environmental concern," hardly the more comprehensive Jackson proposal. The question of what constituted such an area, environmentalists correctly noted, was open to interpretation—a potential loophole for developers. Nevertheless, the proposal included sanctions for noncompliance and represented a tremendous improvement upon the status quo, prompting Train to defend it against charges of inadequacy. When the *New York Times* ran an editorial supporting mandated state-wide plans, Train quickly replied. "We do not want to force all states to take more control from their local governments than is absolutely necessary," Train wrote, a refrain conservatives undoubtedly appreciated.[35]

With his own bill stalled in Congress, Jackson began negotiations with the White House. If they might agree, he assumed, some form of land use legislation had a greater chance of passage. To back up their position, Train and CEQ hired consultants whose subsequent report denied the need for more comprehensive or forceful policies.[36] Recognizing both the administration's concerns and the conservative reality in Congress, Jackson ultimately acquiesced, agreeing to the

CEQ proposal with only minor changes. Train obviously welcomed the agreement, which removed from him the uncomfortable position of opposing the environmental community again.[37] Once again he had reason for optimism. Adding to his hopes, the White House appointed him to a new Domestic Council subcommittee to investigate a "national growth policy." For several years Nixon had advocated such a policy to "influence the course of urban settlement and growth so as positively to affect the quality of American life."[38] Recognizing that although this debate was broader it still involved many of the same issues, Train hoped to use his new position to further the cause of land use.

Unfortunately, by 1973 such was not the case. While Congress added significant funds to the Jackson-Nixon bill, conservatives in Congress grew emboldened. Fearful of infringing upon individual property rights, they rallied around an alternative proposal that removed all requirements that states regulate even areas of ecological concern. With no guidelines or penalties, the bill represented to Whitaker a "no strings attached categorical grant."[39] Train and environmentalists recognized the new bill as a way to emasculate real land use planning but could not convince Nixon to join them in protest. In fact, continuing his tack to the political right, Nixon reversed himself. He embraced the new proposal, effectively ending any real hope for adequate reform.[40] In a similar vein, attempts to formulate a national growth policy floundered. In hopes of reviving the effort, Train asked to lead the study. The reply was as predictable as the death of land use legislation.[41] Train was no longer Nixon's subordinate of choice, land use or any other environmental issue no longer being a priority.

Train had other proposals to protect America's land but, given the new political reality, they fared no better. As a former tax attorney, he well knew the power of the tax code and he advocated revisions to protect historic structures and coastal wetlands. The need was real. Developers had destroyed almost half of the twelve thousand buildings that the Historic American Survey had designated since 1934. In the quarter-century since the conclusion of World War II, America had lost almost 10 percent of her wetlands, and the rate of development had been increasing in recent years. Train's proposal included tax credits for restoration of historic structures and penalties for demolition. Similarly, developers would suffer a tax burden for draining or filling in wetlands. Nixon accepted the idea and proposed it as the Environmental Protection Act, only to have the proposal languish in Congress. With conservatives concerned once again about an infringement upon property rights and never keen on using the tax code to encourage behavior, the bill never had a chance of passage. Even Train hesitated to push too hard, acknowledging at one point to Whitaker, "I, myself, have never been enthusiastic about tax incentives."[42]

Train had more success in his attempts to protect public lands, although even in this regard the success of earlier days largely eluded him. Ranchers grazing their livestock on public lands frequently employed chemical poisons to control coyotes and other predators. Train had a record of opposition to pesticides and quickly recognized that more than just the target species suffered. Frequently hawks and eagles, among others, died. Challenging the powerful ranching lobby, Train pressed the issue with the White House, prompting William Poage, chairman of the House Committee on Agriculture, to respond angrily, "You love coyotes more than sheep."[43] In conjunction with the Department of Interior, Train formed an advisory committee, which promptly took up the cause, ultimately convincing Nixon to issue an executive order banning the practice. The order was a victory for environmentalists although, perhaps predictably, opponents found a loophole to mitigate restrictions. Nixon's order allowed for exemptions in emergencies, which soon prompted the Fish and Wildlife Service to allow use of the M-44, a spring-loaded, cyanide-ejecting tube placed in the ground with a scented bait.[44]

Just as Nixon issued the executive order on predator control early in 1972, he followed with another order prohibiting the use of off-road vehicles. Train and his colleagues at CEQ had pressed for months for such an order, arguing that the rising use of motorized recreational vehicles on public lands disrupted wildlife and damaged soil, watersheds, and vegetation. Train once again returned to the political ramifications, hoping to sway Nixon as he had in the past. Failing to act, Train argued, would leave the initiative to the Democrats in Congress and "could be embarrassing to the administration."[45] Although Train did not make the case, Nixon saw the order as a way to soothe ranchers angry over his actions on predator control. Off-road vehicles disturbed livestock as well as wildlife, and the order was a rare chance to please both ranchers and environmentalists.[46]

Ranchers, of course, were not the only ones interested in public lands; the mining and timber industries dominated much of the economy. With new, large earth-moving equipment, strip miners left much of the public lands a desolate wasteland. Permits to strip federal land doubled in 1970 as the industry turned lush landscapes into permanent pits, polluted the air with sulfuric acid from oxidized coal, and contaminated streams with chemical-laden overburden. Most environmentalists supported legislation to ban the practice but Train reflected the administration's position that any moratorium was impractical. Angering many environmentalists, Train testified in favor of an alternative proposal stressing regulation. Such legislation would allow strip mining but require the restoration of the original topography, including vegetation and water. "A complete ban might be something to work for in the future," Train argued, "but we are in the midst of a very difficult period where energy resources are concerned." Even as most environmentalists protested his bill, Nixon effectively abandoned the proposal as his drift to the political right continued. Train kept pushing but the White House would no longer listen. Nixon's bill was better than nothing, but nothing was the ultimate result.[47]

Clear-cutting was the timber industry's equivalent to strip mining. To his credit, Train forcibly advocated a strong bill to restrict severely the practice in the nation's forests. America faced a lumber shortage and the timber lobby increasingly held Nixon's ear. Restrictions such as Train advocated, the lobby claimed, would drive up costs and hurt the economy. As the Sierra Club filed a lawsuit under the Wilderness Act and NEPA, Train advocated from within, albeit with an equal lack of success. The *New York Times* editorialized that Train's "laudable effort" had "suffered an embarrassingly quick death."[48] Train and his colleagues at CEQ did not welcome publicity surrounding their growing impotence but, in this regard at least, no one could doubt their efforts.

Environmentalists were understandably angered at the administration's growing resistance to regulation, even if their occasional criticism of Train did not always consider his difficult situation. Train certainly could have done more. Although he quietly noted that legislation to increase the timber harvest on public lands was "bad practice environmentally," he remained largely quiet about the administration's poor compliance with the Wilderness Act.[49] He privately believed that the Public Land Law Review report, which encouraged opening up federal lands, "should be thoroughly debated." He did not, however, stress its environmental failings in public even as debate continued throughout his CEQ term.[50] In one sense, such stances deserve a degree of criticism. In another, however, public lands were not his primary responsibility but rather that of the Department of Interior. He had a role to play as head of CEQ but, in fact, he was still part of a team.

If Nixon's domestic priorities shifted, frustrating Train, his interest in foreign affairs never wavered. Nixon still fancied himself a statesman extraordinaire, a diplomatic sage. Train likewise still considered world affairs his own forte, his trips to Japan and Spain only whetting his appetite. This was more than a happy coincidence for the CEQ chair. It presented new opportunities to advance environmental protection even as he found his domestic efforts increasingly restrained. In the President's view, environmental advocacy could advance his policy of detente. It also could burnish the administration's environmental credentials without burdensome regulations or alienating its conservative base. Senator Gordon Allott once told Nixon that environmentalism abroad won no votes at home. Nixon quickly corrected him; it was a relatively painless road with real domestic political benefits. The result was a tremendous success for Train—and the environment. In this regard at least, it was still easy to be both a Republican and an environmentalist.[51]

Just before Train came to CEQ, U.N. Secretary General Maurice Strong proposed a U.N. Conference on the Human Environment for June 1972 in Stockholm, Sweden. Strong found a committed supporter in Train. When several cabinet secretaries initially voiced reservations, Train expressed "concerns about our government's attitude." The United States, Train wrote, should "develop a positive attitude toward the 1972 conference and do everything possible to contribute to its eventual success." Dispatching colleague Gordon MacDonald to Asia for consultation, Train worked with Henry Kissinger to "develop sharp and substantive proposals."[52] Late in 1971, Train met with Strong to iron out details. By early 1972, he testified several times before Congress arguing that the preparation alone had already encouraged nations to work together. "Although it has not yet been held," Train stated, "[the conference] can already be counted a success."[53] Train did not mention the problems he had encountered. The initial list of delegates was "very deficient," he insisted, lacking technical expertise and experience in international environmental affairs. When suggestions arose that a cabinet secretary chair the delegation, Train became defensive. "It is very troublesome," he wrote Whitaker, reciting his international resume. Given his efforts in the preparation, it "would indeed be personally embarrassing to the point that I would necessarily have to consider whether I should go to Stockholm at all." While one might characterize such a reaction as childish, it was effective. Ehrlichman agreed that Train would serve as chair.[54]

Problems persisted. The Department of Commerce complained that Train was acting unilaterally, leaving them out of substantive discussions. Now regularly meeting with the formal delegation, Train worried about reports of "something called the People's Lobby," a group of protesters angry over the Vietnam War. "Frankly," Train wrote, "I am concerned that the whole operation may take on a strong political tinge which could undermine the entire conference."[55] Days before the conference began, Train received "formal instructions to guide you" from the State Department. If nothing else, it was a reminder that, chairman or not, he was still part of a team.[56]

The conference lasted from June 5 to 16. Held in a large convention hall with classical design, much of the conference involved committee meetings held in auxiliary rooms. Delegates from 114 nations attended, along with 400 reporters and representatives of nongovernmental organizations. This latter group ensured there would be a number of rallies, including a "whale march" complete with a scale model of the animal. Interpreters were omnipresent, the perpetual sunlight of the Swedish summer adding to the conference's uniqueness. There was something for everybody, including the presence of former actress Shirley Temple Black, who as special assistant to CEQ, joined the American delegation. While she alone drew coverage, Train developed a friendship based more on her legitimate concern for the environment than the publicity she brought to the cause. George Bush, then at the United Nations, recognized as much and asked Train if he could "spare her for a few days" to help with his own diplomacy.[57]

Aileen and the children joined Train, sightseeing, sailing with friends, and enjoying the show. Train, meanwhile, got down to business. At the welcoming ceremony, Swedish Prime Minister Olaf Palme surprised Train with a sharp rebuke of the United States. Obviously playing to his own domestic audience angered by the Vietnam War, Palme described America's bombing and use of defoliants as "ecocide, which requires urgent international attention." Train was befuddled, but his diplomatic common sense overcame his anger and prevented him from responding in kind with his own opening remarks. Following his instructions, Train contacted Ehrlichman, who agreed that Train should call a press conference and calmly reply. "The United States," Train stated, "strongly objects to what it considers a gratuitous politicizing of our environmental discussions." That night Train and Aileen met Palme in a receiving line, all exchanging pleasantries with tensions obviously cooled.[58]

If this were not enough, oversized egos complicated Train's task. In the middle of the conference Ehrlichman arrived during one of the delegation's early morning meetings. Rather than join the substantive discussion, he immediately launched a loud tirade over his lack of a car and driver, embarrassing all present. Christian Herter, a Train friend and State Department representative who made the arrangements, suffered the brunt of the outburst. "He raised holy hell with me," Herter recalled. Train tried to continue with the subject at hand, disappointed that the highest-ranking American official there cared more about his personal perks than their collective efforts.[59]

The delegation's efforts paid dividends, no thanks to Ehrlichman. Despite ideological, economic, religious, and political differences, delegates agreed on a declaration of twenty-six environmental principles and an "action plan" to implement them. The delegates embraced a number of American proposals, including an ocean dumping agreement similar to the one Train proposed the previous year—then still under consideration in Congress—and an agreement to regulate trade in endangered species. The conference endorsed a ten-year moratorium on commercial whale hunting and an environmental trust fund to help pay for international research and development, the latter also a brainchild of Train's. Overriding the objections of Strong, the conference created a new organization, the Environmental Secretariat, located in Nairobi, Kenya. It also endorsed a number of conservation conventions, including Train's World Heritage Trust.[60] "The United States," Train dutifully reported to Nixon, "played a strong role and gained practically all of its objectives." Train joined Whitaker in arguing for additional publicity, once again stressing the politics of the impending 1972 election. The administration, he claimed, "could capitalize on the momentum developed at Stockholm." Nixon agreed and a press conference followed.[61]

Everyone appeared to agree; the conference was a huge success. "Your contribution was outstanding," Strong wrote Train, a sentiment Train's fellow delegate Senator Claiborne Pell shared. "I thought you did a perfectly outstanding job as chairman," Pell wrote Train. Delegate Norman Livermore wrote Nixon to offer his praise. "May I also take this opportunity to pass along my high commendation for the way Russell Train directed the delegation," he wrote. "The tact, good humor, and efficiency with which he corralled us every morning for ten days and sent us out to our duties made the session particularly enjoyable." Days after the conference, Congress inserted praise into the *Congressional Record*, noting Train by name.[62]

Train and his colleagues deserved the accolades, although admirers overlooked where provisions fell short. China refused to ban nuclear atmospheric testing, while Japan and Norway rejected the whale moratorium. Train had proposed that the United States spend $100 million for the environmental trust fund but the administration reduced the American contribution to $40 million, which by all accounts was only a small percentage of the total funds required. Tensions existed between developed and undeveloped countries as the United

States resisted the latter's attempts at economic compensation. The conference was hardly perfect but, as the world's first major international meeting on the environment, at least highlighted the problem. It began efforts at the necessary cooperation. This alone made the conference a success and Train and his colleagues worthy of the praise they received.

After the conference, Train and his family vacationed in Majorca for a week before returning home. He did not stay long, however, flying to London one week later for a meeting of the International Whaling Commission (IWC). Train wanted the group of whaling nations to adopt formally the commercial whaling moratorium endorsed at Stockholm. As the President's personal envoy, he enjoyed the use of a driver and a large white Jaguar sedan dubbed Moby Dick. Fortunately for Train, Ehrlichman was not there to complain. Unfortunately, despite Nixon's support, his efforts proved fruitless. The IWC rejected Train's efforts for a public vote and then, more importantly, failed to win the two-thirds majority necessary for passage. It was one thing to pass a moratorium at Stockholm, quite another actually to implement one with key countries' economies at stake. Despite his failure, the *New York Times* lauded him for his efforts.[63]

Two months later Train was in London again to implement another of Stockholm's resolutions. This time Moby Dick was gone but Train's efforts bore fruit. Held at the Lancaster House, the conference created the Ocean Dumping Convention to regulate dumping in international waters. This may have stimulated Congress, which finally passed the Marine Protection, Research and Sanctuaries Act of 1972, the ocean dumping legislation that Train had proposed almost two years before. Like the international agreement, the American legislation was a victory for the environment, although not completely warranting the praise Train gave it publicly. While the final bill created the EPA permit program that Train and CEQ originally proposed, it did not specifically ban municipal solid wastes and it allowed COE to control the dumping of dredge spoils.[64]

To advance Stockholm's resolution on endangered species, Train did not have to travel. In January 1973, three months after the Lancaster Conference, Washington hosted a similar conference to address the issue. Train once again led the American delegation, undoubtedly a labor of love given his experience with the AWLF. In fact, representing Kenya was Perez Olindo, the AWLF's first beneficiary. Trade in such products as elephant ivory, animal skins, and exotic pets threatened extinction of a variety of rare animal and plant species. Train won Nixon's support and, with the help of such delegates as Olindo, the conference created the Convention on International Trade in Endangered Species of Wild Fauna and Flora (CITES). Each signatory nation agreed to appoint a scientific authority to ascertain population levels and oversee national regulation. International trade now required approved export and import permits.[65]

Stockholm and its attendant conferences only complicated Train's regular international itinerary. Since agreeing to lead the American delegation to NATO's Committee on the Challenges of Modern Society (CCMS), Train annually traveled to Brussels for the group's plenary session. There he promised to carry on the "Moynihan tradition"—the hard work of his predecessor, Daniel Patrick Moynihan, in organizing CCMS's many "pilot projects."[66] By the time Train left CEQ, CCMS had fourteen such projects, each comprising a group of nations cooperating on an environmental problem of mutual concern. They shared technology and knowledge on such matters as air and water pollution, disaster assistance and road safety. Train made sure his superiors knew of his progress, reporting that "the attitude of the allies is generally good," although it was important to "keep pushing the Germans to play a more active role." Nixon loved the cooperation with his western allies, welcoming the good news even if Train's memoranda bordered on self-congratulation. "Thank you for bringing to my attention the activities of [CCMS]," Nixon wrote. "I am deeply encouraged by the progress you have made."[67] The President, Train recommended, should express his personal appreciation to the countries involved. When traveling, Train routinely stayed at the residence of the American NATO ambassador, first Robert Ellsworth and later Donald Rumsfeld. Even with a full slate of conferences, he still loved to travel. "I indicated to [Ellsworth]," Train reported, "that I would want to visit various key NATO capitals in due course as time made possible."[68]

The cooperation of one country, of course, was critical—an industrial giant not present at Stockholm or Brussels and one on which Nixon obsessed. The Soviet Union was an environmental disaster. Water pollution drained almost $7 billion annually from the Soviet economy with soil erosion adding another $5 billion. Almost two-thirds of the land had some degree of damage. While highly localized, air pollution was a significant health risk in highly industrialized towns. Newly formed governmental bodies, subordinate in ministries dedicated to increasing industrial and agricultural production, were ineffective.[69]

The Soviets recognized the problem and Train recognized an opportunity. Nixon sought to advance his policy of detente and the environment represented an area of common concern. Several lower-level delegations and private organizations had corresponded but no significant cooperation existed. When the

White House announced late in 1971 that Nixon planned a historic trip to Moscow the following year, Train wrote Kissinger and the State Department that an environmental agreement announced on the trip would help make it a success and win positive publicity. The White House concurred and appointed Train the head of an interagency task force to investigate. The group concluded that the Soviets recognized American success in the field and sought to capitalize off it, having decided that no body of Marxist doctrine on the subject existed. Areas for potential cooperation included climate control, earthquake prediction, arctic protection, and maritime and agricultural pollution, among others. The administration, the task force recommended, should exchange scientists and organize bilateral conferences dedicated to implementing joint programs and projects.[70]

After discussions with the White House, Train took Soviet Ambassador Anatoly Dobrynin to lunch at the Metropolitan Club. Train explained that the administration proposed cooperation in eleven areas, ranging from air pollution to the protection of ecological systems. A joint commission, Train recommended, should regularly meet to guide the process. He noted that neither country benefited from arguments over which economic system was the worst offender and that both should stress that "capitalists and socialists are in the same boat and will survive or flounder together." Employing his diplomatic skills, Train frankly admitted that Nixon needed "tangible visible evidence of progress under the agreement prior to the election." The conversation continued as Dobrynin struggled to follow Train's lead and eat soft-shell crabs, which he had never tasted before. "The picture is indelibly fixed in my mind," Train later wrote, "of the distinguished ambassador with almost an entire crab in his mouth, the legs protruding out to either side and wiggling as he tried to chew and speak."[71]

In the weeks that followed, more formal discussions commenced, Train meeting again with Dobrynin and CEQ colleague MacDonald leading a small group to Moscow to draft the agreement's final text. The Soviets demanded only minor, nonsubstantive changes, which Train assumed allowed them to report back home that they had engaged in tough negotiations. Finally, on May 23, 1972, in Moscow, Nixon and Chairman of the Presidium of the Supreme Soviet, Nikolai Podgorny, signed the US-USSR Agreement on Cooperation in the Field of Environmental Protection. Back in Washington an obviously thrilled Train reported to the White House press, "It's a whole new ballgame with the Soviet Union."[72]

For Train, the ballgame had only just begun. Nixon quickly appointed Train to lead the American contingent of the joint committee, naming him cochair with Soviet academician E. K. Federov. To raise publicity and as a gesture of goodwill, Train suggested a gift of two purebred Przewalski's horses then in the Boston zoo. The striking animals, yellow-brown in color and with a black mane and long tail, were indigenous to Mongolia and the central Soviet Union. Few now remained, however, most of mixed breed. The Soviets hoped to reintroduce purebreds to their original range, Train explained, recommending a press release and a photo opportunity with Nixon. The White House thought publicity wise but rejected the gift of horses. It would, Ehrlichman concluded, attract too much attention to Train and detract from Nixon.[73]

The photo opportunity with Nixon the day before the delegation's departure proceeded smoothly. With cameras clicking, Train presented Nixon with an Eskimo carving. Nixon, enjoying his role as elder statesman, explained that the Soviets had an inferiority complex and that the delegation should constantly praise them. It was a good idea, he went on, "to do a lot of sightseeing because the Soviets really want to show things off." In addition to the official itinerary, Train replied, he planned to visit several other cities. Train also noted that his delegation was quite large and included Shirley Temple Black. "Of course, everyone wants to go," Nixon stated, adding that Ms. Black should help garner publicity.[74]

Arriving in Moscow, Train's diplomatic skills were obvious. He thanked his hosts for the "warmth of our welcome" and praised them for their willingness to cooperate. Discussions yielded a number of joint projects: to control air pollution using St. Louis and Leningrad as models; to control water pollution using Lake Tahoe and Lake Baikal; to advance urban design using Reston, Virginia, and the Arctic mining center of Norilsk; and to improve earthquake prediction using California and Tadzhikistan. The two countries agreed to joint studies of oil pollution, wind erosion, pesticides, animal conservation, and forestry, among others. The final twenty-page memorandum of implementation was, according to Federov, "a great beginning to be followed by active work for the benefit of both countries."[75] Not all was work, because Federov and the Soviets proved as diplomatic as Train, entertaining their guests with trips to the Kremlin, several cathedrals and monasteries, the Bolshoi Ballet, art and picture galleries, and even a fashion show.

Highlighting the trip was a visit to the Supreme Soviet, where in addition to Podgorny, Train watched Nikita Khruschev and Leonid Brezhnev preside over a discussion of environmental protection. Sitting in a small box with his colleagues, similar to old church pews, Train watched what he assumed was an event staged for their benefit. In what was one of the few breaks from diplomacy, one Soviet scientist declared, "Of course, socialism by definition cannot pollute." Like Palme's criticism at Stockholm, Train knew that he should not overreact.

Never missing an opportunity to travel, Aileen once again accompanied her husband. Upon completion of the official Moscow summit, the pair joined several other members of the delegation, including Ms. Black, in visiting Leningrad and Irkutsk. Traveling by train in an old paneled car, they visited the famous royal palaces and a bleak World War II cemetery. At a lavish gala on board an oceanographic research vessel docked in Leningrad harbor, Train experienced what must have seemed déjà vu. Just as in Japan several years before, his hosts sang a song and expected a reciprocal performance. Turning to the one person with more singing experience, Ms. Black, Train convinced her to join him in singing "The Good Ship Lollipop." Train once again concluded, "American honor was upheld." In Irkutsk, Train visited an infamous waste treatment plant on Lake Baikal that the Soviets claimed was much improved. Unlike Nixon, who refused to drink treated water from a Chicago plant in 1969, Train accepted his host's invitation for a glass. No one, it seemed, could doubt Train's commitment to diplomacy. Upon his return, Train wrote letters of thanks. "We enjoyed your company, we respect both your singing and your drinking ability, but most of all we thank you for your share in the warm hospitality that we appreciated so much," he wrote.[76]

Once back in Washington, Train met with Nixon again in an attempt to garner publicity. The first session of the joint committee was a tremendous success, Train correctly reported, and he looked forward to the regular meetings scheduled for the future. He had only one complaint that he hoped to resolve. The Soviets did not allow the American press to accompany him outside of Moscow. Given future exchanges, Train reported to Kissinger, "I believe thought should be given now to a coordinated policy on this matter."[77]

In the years that followed, Train became a regular visitor to Moscow, in turn hosting the Soviets when they visited Washington. Train developed a friendship with Federov's successor, Yuri Izrael, at one point hosting a party in his honor, a small taste of the Washington insider culture. Train escorted both Federov and Izrael into the Oval Office, again in attempts to raise publicity. Prior to one such visit, Federov visited the Senate Commerce Committee, much as Train had earlier visited the Supreme Soviet. "I hope academician Federov is more influential with the committee than I usually am," Nixon quipped. Clearly Nixon thought such visits crucial, not so much for environmental protection as world peace. "We must not allow these two great peoples to come into conflict," he commented.[78]

Foreign affairs had become almost a respite for Nixon by the time the 1972 election approached. All around him politics swirled, Democrats angered over his drift rightward and, soon, the press pushing an annoying break-in at the Democratic National Headquarters in the Watergate apartment complex. Although his political strategy appeared to be working and he remained strong in the polls, a growing sense of anger, insecurity, and paranoia consumed the President. He scoured his "enemies list," authorizing a number of illegal activities to secure his reelection. With the publicity of his Oval Office visits, Train did not notice the dark underbelly of Nixon's personality emerging. Earlier, Aileen had visited Katharine Graham when *Washington Post* managing editor Benjamin Bradlee called about publishing the infamous Pentagon Papers. Graham gave the approval and returned to her friends, Aileen unaware of both Nixon's anger and the illegal activities that followed.[79] Secure in his faith in bipartisanship, Train tried to avoid the vicious side of politics. Unfortunately, he was not completely successful.

As the election year dawned, the White House informed Train that CEQ should develop a television special, perhaps hiring one of the popular network reporters as host. Train dutifully began making preparations, which included interviewing potential candidates. Months before the election, however, the *New York Times* called and asked if CEQ were working on such a special and had considered CBS newsman Daniel Schorr. Train knew immediately that the White House had duped him. Recent news reports claimed that the FBI had launched an unethical investigation of the liberal Schorr, a frequent Nixon critic. By answering truthfully—yes, he was interviewing candidates for such a show and, no, Schorr had not been interviewed—Train had unintentionally provided the White House an excuse. The White House could claim the investigation necessary for Schorr to accept the CEQ job. Train was more ashamed than angry but swallowed his pride. If he wanted to remain part of the team, he could not go public with the truth. It was not easy. Schorr cornered him at a party and asked what really happened. Ruckelshaus stepped forward and provided his colleague an escape. "Maybe they were going to offer you my job," he wryly noted. His eye on the big picture, Train remained quiet.[80]

The frustration grew. Just as the Schorr incident took place, the White House attempted to delete several sections of the CEQ annual report. Train and his colleagues drafted a report that adequately reflected the environmental reality—

progress in air pollution but continuing problems in most other regards, with a total cost approaching $287 billion. Faced with conclusions that ran counter to its new agenda, the White House blundered. "The trick," Whitaker privately wrote, "is to snuff out some of the chapters which are causing problems early in the process." Unfortunately for the administration, the *Washington Star*'s Roberta Hornig discovered the ploy and publicized it. Democratic presidential candidate George McGovern pounced, declaring Nixon's "suppression" of the document "absurd." The glare understandably turned on CEQ. Rather than acknowledge the truth, Train and his colleagues once again remained quiet. As CEQ released the chapters, MacDonald declared them "only working drafts," the deletions not suppression but normal revisions. Suspecting more, much of the press attacked. "A brickbat to Russell Train," the *Hartford Times* editorialized. His report "is marred by a Pollyanna view of progress and censorship of three important areas." The former charge was unfair but the latter right on the mark.[81]

Train pushed ahead, CEQ proposing legislation requiring companies to test all new chemical substances prior to marketing. EPA, Train argued, should restrict any chemical that posed "unreasonable risks." It took time but Congress eventually passed the proposal as the Toxic Substances Control Act (TOSCA).[82] With the election approaching, Nixon signed several bills that CEQ supported, although Train played little role in their passage. Originally proposed by Whitaker's task force, the final legislation significantly improved environmental protection. In each case, however, deliberations weakened provisions. The Federal Environmental Pesticides Control Act created an EPA permit system but, at Nixon's insistence, included significant concessions to farmers. It kept House Agriculture Committee oversight and required an EPA indemnity for financial losses. The Marine Mammal Protection Act instituted a permanent moratorium on the killing and importation of ocean mammals but allowed for Commerce Department–approved exemptions. The Coastal Zone Management Act provided a grant program similar to that proposed for broader land use but, once again, allowed the Commerce Department primary jurisdiction.

Train praised each bill, repeating the administration's line that the legislation proved Nixon's commitment was strong. He was a loyal member of the team even as the frustrations continued to mount. The administration proceeded with testing of a controversial nuclear warhead in Alaska's Amchitka Island, with Train's pleas that the bomb might cause an earthquake or tsunami falling on deaf ears. Fortunately, he was wrong. After including Train's proposal for an Environmental Policy Institute in his 1971 environmental message, Nixon finally abandoned the project. Train argued that he had already interviewed potential directors and pressed to "move this along." In the end, he received no reply.[83]

Several years before, Nixon had accepted Train's advice that a close public alliance with the National Industrial Pollution Control Council (NIPCC) was unwise. Now with the election approaching, Nixon invited its members to the White House and declared that he appreciated their work. When the Commission on Population Growth and the American Future issued its report endorsing sex education, increased funding for family planning and easier access to abortion, Nixon predictably distanced himself. Train recommended an interagency task force to study the commission's conclusions and Ehrlichman agreed. He did so, however, more to stall the issue than change policy. When Train's task force issued its own report largely supporting the earlier commission, the White House ignored it. According to one White House aide, "There is little to be gained from presidential action."[84]

Train was not the only one frustrated. In the middle of the presidential campaign, CEQ colleagues Robert Cahn and Gordon MacDonald resigned. While Cahn avoided direct criticism of Nixon, his disappointment was obvious. "The bureaucracy has grown slow in giving weight to environmental factors in decision-making," he explained. Correspondence with the White House was "less than ideal." MacDonald, Train recalled, "had grown restless." To bid his colleagues farewell, Train hosted a lavish party and invited many of his Washington connections. The site he chose was the early nineteenth-century Decatur House, the original home of Commodore Stephen Decatur, who was famous for declaring, "Our country, right or wrong." To replace the departing duo, Nixon appointed Beatrice Willard and John Busterud, assured that they were "more reasonable and moderate on environmental affairs." Equally troublesome from Train's perspective was the resignation of Whitaker and his replacement by subordinate Richard Fairbanks, a younger man who cared about the environment but was "much less active on the issue."[85]

Whitaker's departure meant that Train lost a key conduit to Nixon. In any event, reaching the President was going to be more difficult, adding to Train's frustrations. Angered because Congress refused to pass his government reorganization, Nixon proposed to accomplish the same thing without congressional approval. His plan called for four "counselors," in essence "super-secretaries" who would funnel information from the other cabinet members. Named Counselor for Natural Resources was Secretary of Agriculture Earl Butz, a man Train concluded was "pretty good at rabble-rousing against the environment." He was a "very articulate guy," Train acknowledged, "but I never agreed with him on much of anything." Now, if he wanted to reach the White House, he had to go through Butz, which was never an easy task.[86]

Fairbanks soon complained that Train, Ruckelshaus, and Secretary of Interior Morton ignored him. Butz replied with a letter to the trio reminding them that he was their "liaison" to Nixon and while they could recommend "topics for the agenda," he was the only avenue of dissent and the final arbiter of their objections. "It was a growing nuisance," Train recalled, "imposing one more layer of bureaucracy between me and the president." In public, however, Train remained the good soldier. Asked about the new counselor arrangement in an interview, he kept a stiff upper lip. "I don't consider that Secretary Butz is between me and the president at all," he declared.[87]

"My relationship with Nixon had become ninety percent ceremonial, just window dressing," Train recalled, obviously for a moment forgetting his role in detente. Any presidential comment that reflected a genuine interest in the environment shocked him. Just before one of the ceremonial photo ops, the presentation of the 1972 White House environmental message, Nixon suddenly turned to Train. "Russ, I want you to tell the people something for me," Nixon began. "Last weekend at Key Biscayne the roads were all clogged and we've got to do something, perhaps more mass transit. Three-fourths of the cars had only one driver." Train could not believe his ears, Nixon's expression of concern unnerving him. "Three-fourths of the cars," Train quoted the President in the subsequent new conference, "had a driver."[88]

Train knew what really worried Nixon—a potential energy shortage, which by mid-1972 appeared a distinct possibility. The *New York Times* predicted a 50 percent increase in oil consumption over the decade, far outstripping domestic production. America's dependence upon oil from the Organization of Petroleum Exporting Countries (OPEC), with whom she had a tenuous relationship, was sure to grow. Such concerns were real but for Nixon required one key solution—weakening environmental regulations. At one point an aide recommended energy conservation. "What's that?" Nixon asked. Rather than increasing alternative fuels, Nixon encouraged Congress to pass legislation exempting nuclear power plants from NEPA. Angry over lawsuits that delayed the Alaska pipeline, he reflected the comments of many automakers who decried air pollution standards.[89]

Train could take it no more. Only months before, he had sent a lengthy memo to Ehrlichman bemoaning "a substantial erosion in the [President's environmental] position." The administration had an "extraordinary record" but increasingly wasted one of its "major assets." Unfortunately, despite appeals to the politics of it all, nothing had come of it, the memo not even warranting a reply. Now, with the election only months away, Train decided to go public. While not naming Nixon or commenting on broader policy, he publicly protested "the current tendency to make the environment the whipping boy of our energy problems." Citing the need for conservation and sacrifice, the prepared speech declared that such conclusions "obscure the facts, confuse the issues, and can only serve to delay effective solutions of our energy policy."[90] Like Cahn's comments, it was not a direct denunciation of his superiors but was a clear break from the White House. In the midst of the political season, the good soldier could no longer completely hide his frustration.

The 1972 election campaign did not make Train's task easier. Early in the year Whitaker implored Train and Ruckelshaus to make public appearances touting the administration's record, in both cases appealing to their egos. While his letter to Ruckelshaus declared the EPA administrator "the best spokesman on the issue," his letter to Train concluded essentially the same thing. "There is no one within the Administration who can take your place as the leading advocate on the environmental issues," Whitaker wrote. "In terms of background, ability and scope of responsibility, you are in the unique situation to present and advocate the President's complete environmental program."[91]

On one hand, Train agreed. The administration needed him and, in his view, its overall record was worthy of praise, its threatening retreat notwithstanding. To laud the White House was, after all, to applaud his own performance. Train indeed had an ego and he knew that criticism of Nixon left himself an ineffective advocate. On the other hand, his commitment to the environment required him to challenge inadequate policy. He did not want to appear an unwitting tool or political cover for a man intent on retreat. Train, therefore, decided to avoid as many press conferences as possible. He would support the administration but not call attention to himself. While he agreed that Congress deserved as much blame as the administration, neither would he criticize Nixon's opponents. The White House's desire for such was "counterproductive and ineffective," Train assumed, reflecting once again his faith in bipartisanship. Train did not mind appealing to Nixon's political instincts in private but was "uncomfortable playing such an overt role politically." It was a fine line that he had trouble walking.[92]

Even as his frustrations grew, Train joined his administration colleagues on a chartered jet to Miami for the Republican National Convention. Along with Nixon and the others, he and Aileen stayed at the Doral Beach Hotel. He rode with George Bush, now director of the Central Intelligence Agency, in a CIA limousine to the convention hall. He also joined Ruckelshaus and Morton testifying before the Platform Committee, all arguing for recognition of the problem and a strong plank in response. In such friendly confines, Train did not hesitate to praise Nixon's "significant advances." The White House's record, Train concluded, "is not one of speeches and promises but one of action and accomplishment."[93]

In the end, the environment played little role in Nixon's landslide election, the Vietnam War and the economy once again overshadowing most other issues. Nevertheless, both campaigns stressed environmental protection much more than four years before. Perhaps as a reflection of Train's own testimony, the Republican National Committee devoted more attention to the environment than any other domestic issue. Train felt that he did not have to risk his or CEQ's reputation; Nixon's campaign made the message perfectly clear.[94]

The campaign over, the environmental rhetoric faded and, for Train, it was back to the new reality. The dark underbelly of Nixon's personality was now out in front for all to see. Just after the election, Nixon surprised his staff with demands for their resignations. This was more than the pro forma requests for resignations that had characterized earlier presidencies at the beginning of their second term and, despite the loyal efforts of many staffers during the campaign, Nixon fully intended to clean house. Not surprisingly Train was abroad on CEQ business and learned of the order in a telephone call from Aileen. Angry, he immediately went to the American embassy at Grosvenor Square and typed a one-sentence letter of resignation. Once back in Washington, Train could take the wait no longer and telephoned Ehrlichman, who, thankfully, told him he was to stay. Not all were so lucky. Weeks later, Train and Aileen attended a White House dinner. Unlike Train, most guests still did not know their fate. Sitting next to him was the wife of a new Justice Department attorney, who explained that she and her six children had just arrived after selling their San Francisco home. There was no happy ending; the White House soon accepted her husband's resignation. Aileen heard no such sad tales but, rather, had to endure sitting between Ehrlichman and Secretary of State William Rogers, the tension between them "palpable."[95]

The new year brought more bad news—for Train and Nixon. Train learned of significant cuts in his budget, with CEQ's personnel ceiling reduced from sixty-five to fifty. Wrestling with the necessary adjustments and terminations was not easy, adding to the chill all over Washington. Most of the chill, of course, was the Watergate scandal, now slowly creeping into the White House. With his new counselor arrangement only one example, Nixon continued to isolate himself, growing angrier by the day. Step-by-step news reports uncovered top aides' culpability, ultimately ensnaring Ehrlichman and Chief of Staff H. R. Haldeman. Most of the administration's policy—if not government itself—slowed to a crawl.[96]

Never one to rest easy, Train sought to rescue both his cause and career from the growing chaos. Through an old friend, White House aide Bryce Harlow, he arranged an interview with Vice President Spiro Agnew. There he recounted not only his frustrations with environmental policy specifically but the impact of the scandal on the Presidency itself. If the American people were to continue to have faith in the Executive Branch, Train suggested, Nixon should appoint strong leaders in the cabinet and agency posts and then allow them the autonomy to make their own decisions and develop their own political base. Reliance upon a small cadre of White House advisors to the exclusion of everyone else increased instability and helped tar otherwise independent agencies with the sins of a few. Train believed that, like a diverse and healthy ecosystem, Nixon should encourage a broad spectrum of views. Agnew might have taken Train's comments as a power play. Indeed, he told Train that nothing annoyed Nixon more than ambitious staffers trying to fill perceived power vacuums. Train, after all, stood to gain personally. Surprisingly, however, Agnew agreed, frustrated in his own inability to reach Nixon. Too often, Train and Agnew commiserated, junior White House staffers inhibited direct communication, not returning calls and overruling more senior officials—an absolute outrage. Agnew was not optimistic, claiming Haldeman's replacement, General Alexander Haig, was no better. Train left with no solution in hand but with the realization that he was not alone in his frustration.[97]

Train may have agreed with Agnew about Haig but planned to take no chances. "I was delighted to see the great news of your promotion," Train wrote the general. "Congratulations and every success in your new responsibilities."[98] If he could not change the system, he could at least grease the wheels to the best of his ability. Train never saw Agnew again. Soon after their meeting, Agnew resigned over charges of tax fraud. Train and Aileen joined Laurance Rockefeller and his wife, Mary, on a Caribbean cruise, undoubtedly a welcome respite from the political frenzy engulfing the nation's capital.[99] Rockefeller had just tendered his resignation as chairman of the Citizen's Advisory Committee on Environmental Quality and, while not mentioning it, Train must have pondered his own political future. Perhaps like the sun setting over the Caribbean, his own day was coming to an end.

7

A Challenge in Crisis

By the summer of 1973 Watergate was more than just the name of an expensive apartment complex in the nation's capital. The burglary at the Democratic National Committee the previous summer had spawned criminal and Senate investigations, which riveted the public as each new revelation crept higher in the White House hierarchy. Watergate was now synonymous with scandal and dirty politics, a growing governmental paralysis that accompanied a preoccupied bureaucracy. Was Nixon involved? How would a wounded President deal with high fuel costs, a looming energy shortage, and turmoil in the Middle East, the only other issues that seemed consistently to challenge scandal for the headlines? In the heat of the summer, few had reason to smile.[1]

Certainly not Train. Nixon's drift rightward was obvious, the economy and the White House's political difficulties exacerbating the administration's environmental retreat. To defend himself, Nixon sought to solidify his conservative base, denouncing the environmentalists he had earlier courted. Frustrated at every turn, Train might have considered resigning if not for his love of policy and a resilient optimism that he might still advance his cause. Nixon had no plans to give up easily and neither did Train. Their fates seemed intertwined even as the distance between them grew.

Unlike Nixon, however, fate had always been Train's friend, the events of the previous year aside. As if to prove that at least a sliver of his earlier good fortune remained, the very scandal that appeared to hamper his cause offered Train an unexpected opportunity. Following the death of longtime Federal Bureau of Investigation director J. Edgar Hoover, Nixon chose L. Patrick Gray as Hoover's successor. Reports soon implicated Gray in Watergate-related document destruction, however, and Nixon moved quickly to name William Ruckelshaus as the new nominee. Train was happy for his friend and called to congratulate him but he also recognized immediately that the appointment necessitated a new administrator at EPA. He did not hesitate; he wanted the job.[2]

Train had good reason, as he explained to the press in the weeks that followed. The struggle to protect the nation's environment had reached a new phase. With much legislation already in place, the critical battle over enforcement loomed. "EPA is where the action is," Train noted. "Implementation of existing environmental legislation will be the thrust of future federal activities."[3] Train knew as well that EPA, unlike CEQ, was a "mainline, regulatory agency" outside direct White House control. Faced with Nixon's environmental retreat, he might enjoy more independence at EPA and more impact on the environmental quality he valued.

Train had worked to cultivate a good relationship with Alexander Haig as soon as the former general replaced H. R. Haldeman as Nixon's Chief of Staff. Now Train called on his new friend to let him know of his interest before any momentum built for other possible candidates. Haig recognized that Train was popular and that an EPA appointment was sure to win positive publicity, something Nixon desperately needed. The key was to ensure that an overzealous Train not become a thorn in the White House's side. Sitting in Haig's West Wing office, neither man spoke of Nixon's antipathy toward environmentalists. Both understood the situation well, however, and Train offered the requisite promises of a "balanced approach" even as he quietly regarded Nixon's retreat as anything but balanced. Haig reported Train's position to Nixon, with an announcement of Train's nomination quickly following.[4]

The overall reaction was positive, if not the universal acclaim that greeted Train's appointment as CEQ chair. He now had a more extensive record, with the environment a more controversial and divisive issue. "I don't envy Russell Train his appointment as administrator of the U.S. Environmental Protection Agency," columnist Dale McKelvie wrote. "What EPA really needs is a cross between Dracula and King Kong."[5] Threatened by Nixon's former accomplishments, business demanded lax enforcement. Angered by Nixon's new attitude, environmentalists demanded the opposite. "In that job," one environmentalist admitted in a moment of candor, "you're damned if you do and damned if you don't."[6]

Train's Senate confirmation was never in doubt even as the *Wall Street Journal* reported that his nomination "draws fire from both sides."[7] On one side Republican senators Clifford Hansen of Wyoming and William Scott of Virginia provided a conservative critique, placing a hold on the nomination. Train apparently had done too much for the environment. Hansen, all knew, acted at the behest of the coal industry, which was anxious over potential mining restrictions. Scott, however, was no traditional westerner opposed to specific regulations. He represented the new ideological conservative, his opposition to Train more a statement against federal power and, as events would soon prove, a sign of times to come. Train met with Scott several times but could not sway him. "Russell Train appears to be a very fine gentleman, well-educated, cultured and a devoted public servant," Scott told Senate colleagues. "My principal reservation is whether Mr. Train would put undue emphasis on the environment to the detriment of our standard of living." For Scott, it appeared, federal regulation did not ensure adequate living conditions but threatened them. Scott noted that Train described himself as a "conservationist" in his *Who's Who* biography. "Are you ashamed of being a judge?" he asked. Train knew that nothing he might say would make a difference and, uncharacteristically, remained angry for years. In his self-published memoirs, Train noted that Scott called a press conference to deny a press gallery vote naming him the dumbest member of the Senate.[8] Hansen's opposition, on the other hand, hurt Train more than angered him. Hansen avoided Train's efforts to lobby him, even though the two had known each other for years as Washington insiders. "It was odd," Train recalled, obviously reflecting traditional, but fading, political mores. "Generally when you do something like this, you call them up and say, 'Hey, look, no hard feelings . . . just politics.'"[9]

On the other side, some environmentalists complained that Train had not done enough for the environment. A debate raged over whether the government should allow "significant deterioration" of air quality in states already above federal limits. Some reports incorrectly claimed that Train favored the "local option," pleasing conservatives but angering his traditional constituency. Others complained that Train did not fight the Alaska pipeline forcibly enough. "Where was he when we needed him?" one environmentalist asked. "His problem was that he was too tightly tied to the White House, where someone apparently told him to shut up." The most public criticism came from Ralph Nader. The consumer advocate had met Train several times but now declared, "I don't think [Train] cutting edge." Train, Nader concluded, "isn't temperamentally suited for an enforcement role."[10] The environmental community did not resist Train's nomination but, in the words of Brock Evans, the Sierra Club's Washington representative, "Some of us wish he had been more vigorous in publicly supporting environmental causes while at CEQ."[11]

The environmental criticism bothered Train more than the conservative critique, which to an extent all expected. Cognizant that his confirmation was not in doubt and undoubtedly glad to place some distance between himself and the growing scandal at the White House, Train took every opportunity to stress his independence. "I intend to run EPA and make the decisions," he answered when a reporter asked about pressure from the White House. "I will listen to other agencies and the [Office of Management and Budget], but I have a statutory responsibility under the air and water quality laws that cannot be delegated or abrogated." After Hansen and Scott finally lifted the hold on his nomination, Train made the same point in his confirmation hearings. "EPA is an independent agency within the Executive Branch," he began. "It has many regulatory and enforcement functions established by statute. In all such cases I assure you that I, as Administrator, will make the final decisions."[12]

Train lobbied senators and dealt with the press like a seasoned Washington bureaucrat. His staff prepared him for questions and developed "briefing points," while Train corresponded and visited with supporters. In private, the staff surprised him with a book of congratulatory letters. One was a joke, a fictional letter supposedly from Scott. "Mah people don't have nothing against you fancy striped pants shysters," the letter read. "But ah still have a hard time telling the tractor mechanics, irrigators, and bee keepers of mah fine State how the President in his right mind could have picked a left wing, anti free enterprise feller like yourself to run the Environmental Power Commission." The staff could joke, because the press largely came to Train's defense. According to one pundit, Scott was a "Neanderthal." In the end, no one spoke against Train when the Senate began its formal debate. The vote was unanimous, 85–0.[13]

As in the past, Train once again met with Nixon in the Oval Office. The meeting was brief and the two men never sat down, but Train took the opportunity to ingratiate himself with the President. He asked Nixon to give him the oath of office as the new administrator. Train knew that Nixon would decline; rarely did presidents personally swear in appointees and, if he had, it would appear overtly political given the clamor for Nixon's head. "It would be the worst thing I could do for you," Nixon replied, a self-effacing line that he enjoyed saying. Train could praise Nixon, because he believed that he had accomplished all that he wanted. He had stressed his independence publicly and, cognizant that the embattled President needed the popular appointment, had even secured a written promise.[14] He had also secured a written pledge that he would continue to lead the nation's international environmental agenda. "As you assume your important new role as Administrator of the Environmental Protection Agency," Nixon wrote Train, "it is my wish that you continue to carry out responsibilities in the field of international environmental cooperation."[15] It was music to Train's ears.

One might have forgiven Train if he had thought for a moment that his luck had once again turned, that his new position made his growing frustrations a thing of the past. The reality, of course, was different. While EPA was indeed an independent agency that afforded Train a leeway he had not previously enjoyed, it was not completely devoid of politics. As early as 1971, Nixon had pressed Ruckelshaus to include political supporters among many of the deputy administrators. When Kansas Republican Senator Robert Dole had recommended several prominent Republicans, Ehrlichman had agreed, noting that "we are very much in debt to him." This had prompted Whitaker to warn of "negative public reaction that the choices were made on political grounds rather than the technical qualifications of the appointee."[16] With environmentalism then at its zenith, Ruckelshaus had done an excellent job of resisting political pressure. Now, however, energy concerns threatened to eclipse environmentalism and Nixon thought it prudent to warn Train that times were different. Even as Train publicly declared EPA's independence, Nixon was blunt. "This area of concern," he told Train, "was moving into a new and more difficult period of maturity." The earlier "blush of emotion and commitment" had passed and the new task was to "balance environmental protection with the nation's other pressing needs."[17]

The same week that Nixon announced Train's nomination, the *Washington Post* ran a front-page article entitled "Nixon Aides Now Fill Key Agency Jobs." The article noted that while many White House staffers faced Watergate-related legal trouble, others had "infiltrated line agencies" that demanded expertise and independence. A Democratic-driven House subcommittee investigation reported that EPA employed twenty "ex-Presidential aides," second only to the Department of Commerce's twenty-five. Undoubtedly many of these aides possessed the requisite skills and were excellent appointees. Others, however, were purely political. Several had worked for the Committee to Re-Elect the President (CREEP), with one simply listed as "confidential representative to the White House."[18]

The White House exerted pressure on EPA in other ways. As Nixon began to back away from environmental advocacy the previous year, Ehrlichman had launched a new internal administrative procedure, the Quality of Life Review. On the surface the order had simply promised the "proper balance between economic and environmental considerations." In practice, however, EPA had to submit all proposed regulations to other agencies. Coordinating this review was the Office of Management and Budget (OMB), which had already shown itself to be no friend to Ruckelshaus's environmental bureaucrats. Train, who acknowledged the review as a "troublesome process," now faced this pressure like his predecessor. When the *New York Times* incorrectly reported that OMB restrictions had contributed to Ruckelshaus's departure, Senator Muskie called a press conference to denounce White House interference. As Train assumed the helm of EPA, the Quality of Life Review remained in the news for both its supporters and critics.[19] Train's aides recommended that he follow Ruckelshaus's lead, avoiding criticism of the review overall but, in private, challenging specific regulatory conclusions. He might also work to disband the review in any further executive reorganization.[20]

If Train prepared himself to withstand political pressure, he still had more to master. The evolution of environmentalism from policy formation to implementation placed science at the fore. Industry had already learned that the best resistance was to challenge the scientific integrity of standard-setting and it employed a bevy of experts on its behalf. Debate increasingly revolved around highly technical minutiae that only trained scientists might comprehend. Train, however, was no scientist. He wanted politics removed from EPA but, in advocating such, removed his own expertise. According to Laurence Moss, who led the Sierra Club at the time, "[Train] didn't have enough technical knowledge to know if the staff work submitted to him was going in the right or wrong direction." He had good intentions but was "completely dependent upon the data collected by subordinates and their implementation of it."[21]

In a similar vein, Train lacked, in the words of the *Washington Post*, "administrative experience in running a sprawling bureaucracy like the EPA." He had demonstrated his keen interpersonal skills and had effectively managed CEQ. The staff of EPA, however, dwarfed its Executive Branch cousin. In addition to its extensive Washington bureaucracy, EPA operated ten regional headquarters around the nation, which regularly dealt with hundreds of state and local governments. EPA also employed a number of research facilities, providing the data that Train would have to digest. Ruckelshaus had granted the regional offices a significant deal of autonomy, which challenged Train, who was accustomed to CEQ's "strong chairmanship" structure. Ruckelshaus agreed to help in the transition and the two men "laughed about" what faced them.[22] Characteristically, Train brought with him a number of able deputies, including John Quarles and

Alvin Alm, both of whom had worked with him for years. Their assistance, he hoped, would shorten his necessary learning curve.[23]

EPA operated from a converted apartment building near a shopping center in southwest Washington, far removed from most other agencies and departments. Train did not mind losing his prominent view of Lafayette Park and the White House; it only underscored the independence for which he hoped. He also admitted to welcoming the publicity that surrounded his new job; he was in the middle of policy, if not the middle of town. Just after meeting with Nixon, Train took his oath of office before friend and new Attorney General Elliot Richardson, Aileen holding the Bible as the pair stood before a large EPA banner. Nearby stood Ruckelshaus and Robert Fri, who had operated as acting administrator while Train's nomination was on hold. If they were glad to leave their jobs, cognizant of the task before Train, they gave no indication. It was a happy occasion, a respite from the frustrations attendant to Nixon's environmental retreat.

The celebration did not last; Train's luck had not changed. As if a precursor to difficult times ahead, an attack of painful kidney stones hit Train just as he and Aileen returned to their Eastern Shore farm. As he recovered, news arrived that on Saturday morning, October 6, the holy day of Yom Kippur, Egypt and Syria had attacked Israel. Train watched as events unfolded, undoubtedly aware of their implications for his own task. Nixon quickly reinforced the Jewish state, which in turn prompted the Arab countries of OPEC to launch a complete oil embargo against the United States. For months Americans had feared an "energy crisis." Now, however, it was a reality. Gasoline prices skyrocketed and shortages were commonplace. It all meant one thing—a renewed momentum for the relaxation of environmental law.[24]

Complicating matters was a surprising development in Watergate. The summer Senate hearings had uncovered the existence of secret White House tapes, which Nixon's own special prosecutor, Archibald Cox, subpoenaed. When the courts ruled that Nixon had to relinquish the tapes, Nixon ordered Richardson to fire Cox. Richardson refused and resigned, suddenly making Ruckelshaus the new Attorney General.[25] Ruckelshaus also refused and suffered the same fate before Solicitor General Robert Bork carried out Nixon's order. Train watched the infamous "Saturday Night Massacre" in amazement, sad for his friends, who only weeks beforehand had stood by his side. After sending a letter of condolence and support to Ruckelshaus, he left for Brussels and a meeting of NATO's Committee on the Challenges of Modern Society. There he arrived at the house of the NATO ambassador, Donald Rumsfeld. Rumsfeld answered the door and asked of news from home. When Train informed him of the firings, the stunned Rumsfeld's "jaw dropped."[26] The energy crisis and Watergate, all knew, seemed far from abating.

When Train received an invitation to attend one of the regular cabinet meetings, he knew what to expect. The President invited him "because we have some difficult decisions to make with regard to the environment." As if to soften any resistance, Nixon lauded him as a "team player." Energy Policy Office (EPO) director John Love began the discussion by noting problems inherent in the Clean Air Act. Train had always been a strong supporter of the law, which at the insistence of the Senate mandated stringent auto emissions standards, a 90 percent reduction in hydrocarbons and carbon monoxide by 1975 and a 90 percent reduction in nitrogen oxide by the following year. The law empowered EPA to set national ambient air quality standards and mandated specific emissions limits for sulfur oxide from stationary sources. It permitted EPA to grant temporary emissions variances but, as federal courts soon ruled, it prohibited deterioration in the air quality of states already above federal requirements. After listening to Love, Nixon declared it necessary to "lower the emission limits" because the shortage of oil required greater use of sulfur oxide–producing coal. Train objected and noted the ramifications for human health. Nixon remained unmoved, however, remarking that in his youth more cases of tuberculosis arose from cold homes than any other cause. Recognizing that he was losing the argument, Train received promises that the White House only sought the temporary variances already allowed, not a permanent weakening of the law's standards. Train did not help himself with an involuntary scream of pain, the result of an attack of sciatica that hit just as Nixon spoke. Haig, seated across the table, stared in wonderment as a startled Nixon went on with the discussion. Soon after the meeting adjourned, Nixon and Love went before the press to break their promise, calling for a permanent relaxation of air pollution standards. Although this was obviously duplicitous, Train granted Nixon the benefit of the doubt. "The presidency was collapsing around him," he recalled, and the issue was extremely complicated.[27]

Indeed, the issue's complexity matched its importance. Just before Train arrived at EPA, Ruckelshaus denied the auto industry's application for a one-year variance in the hydrocarbon and carbon monoxide standards, in essence tying the industry into experimental catalytic converter technology, the only hope to meet the required standards. Unfortunately for Train, his arrival coincided with news that the catalyst technology produced small amounts of sulfuric acid vapor, adding a new dimension to the debate. Many EPA staffers worried that Train's White House connections and lack of scientific background might weaken his resolve. "He wanted to get into the saddle fast," according to one employee, "and tried too hard to impress." Aware of Train's genial nature and reflecting the concerns of Nader, another staffer, Robert Sansom, openly questioned whether Train had Ruckelshaus's backbone. "The question," stated Sansom, "is will he know where to draw the line on compromises?"[28]

Sansom soon got his answer. With his own staff divided on the issue of catalyst technology and the debate "sometimes heated" within his own office, Train concluded that abandoning the technology and, thus, the permanent standards, posed a greater risk than the sulfuric acid vapors. Testifying before the Senate Public Works Committee, Train acknowledged that the data was uncertain, in effect leaving the onus on critics of the catalytic converter to prove their claims beyond doubt.[29] He did not leave out the possibility of granting temporary variances but, undoubtedly pleasing his EPA subordinates, remained firm on the law's permanent standards.

Back in the cabinet, Love pressed his case. Given the Arab embargo, he argued, the administration should propose legislation to exempt all energy facilities from environmental restrictions. Train thought the proposal "extreme" but Nixon listened intently. The following day Nixon went on national television to announce "Project Independence," his program to make the nation energy self-sufficient by 1980. The address referenced conservation, including proposals for a maximum fifty-five-mile-per-hour highway speed limit and the continuation of daylight savings time. The thrust of the program reflected the arguments of Love more than Train, however, posing a genuine threat to legitimate environmental protection. Risking White House anger, Train was publicly skeptical. "I doubt that we can achieve full independence of all external sources by 1980," he declared, "with or without substantial damage to the nation's environment."[30]

It was the beginning of a major public rift between Train and his superiors at 1600 Pennsylvania Avenue. In early 1974, with the embargo now over but oil prices still above $11 dollars per barrel, Nixon ordered OMB and the Federal Energy Office (FEO, a new replacement to EPO) to craft specific proposals to emasculate the Clean Air Act. For Train, having EPA ordered to provide data but essentially left out of deliberations was a worst-case scenario. OMB and FEO proposed weakening the Clean Air Act's permanent hydrocarbon, carbon monoxide, and nitrogen oxide standards as well as extending deadlines for compliance. Train did not object to further temporary variances but did to a proposal requiring EPA to consider economic as well as health aspects in fulfilling its statutory obligations. In regard to sulfur oxide emissions from stationary sources, the proposals essentially abandoned the use of smokestack "scrubbers," the best alternative to fight pollution but also the most expensive for the coal industry. The industry sought the permanent use of tall smokestacks, which did not reduce emissions but better dispersed them, allowing for legal compliance. Train, aware that industry had to cut power production when weather and wind conditions demanded, did not object to this practice as a temporary measure. He could not, however, accept this "pollution dilution"—a catchy phrase he coined to garner attention—as a permanent solution.[31] OMB and FEO also proposed the preemption of federal law over more rigid state requirements and allowed the deterioration of air quality already cleaner than federal mandates, a controversial proposal the courts had just held contrary to the law. Nixon was already on record supporting legislation to exempt the Alaska pipeline from NEPA's impact statement requirement, the pipeline still in contentious litigation, but now OMB and FEO proposed to exempt all energy-related activities as Love had originally advocated.[32]

It was all too much for Train, who continued to object publicly. It was an unusual development, not only in the sense that Train had always been the "team player" that Nixon described but also simply for the fact that intra-administration disputes rarely saw the light of day. More often than not agencies and departments resolved their disagreements in private before a formal proposal reached Capitol Hill. Train, however, declared that he would "fight against [the proposals] to the last wire." More significantly, he would not testify before Congress in support of them. Some reports claimed that he threatened to resign, although Train denied it. The media pounced on the dispute, at least to the extent that anything could draw attention from Watergate. For Train, much was positive. "We desperately need more people in and out of government with the integrity and the courage to face reality and do what is right," declared *Motor* magazine. Some conservative outlets were less kind. The *Wall Street Journal* editorialized that Train resisted all revisions of the Clean Air Act, an incorrect statement that Train addressed with a letter to the editor.[33]

With the debate raging, the White House asked Train to work out his differences with OMB and FEO. The result was agreement on eleven of thirteen proposals. Train convinced OMB and FEO to drop the NEPA exemption for all energy facilities and the mandate for EPA to consider economic costs.

Train agreed to extend auto emissions standards for two years and to the additional use of coal. Neither side, however, wanted to compromise on the issue of tall stacks or the deterioration of air quality above limits. The administration, therefore, presented a divided front to Congress, which was now left to resolve the issue.[34]

With the energy crisis as a backdrop, Congress finally produced in June the Energy Supply and Environmental Coordination Act of 1974. Although some of the administration's proposals met stiff resistance, high fuel costs scared the public and, consequently, the new legislation reflected many of the compromises Train had reluctantly accepted. The law granted auto manufacturers "two additional years of grace for the development of emission control technology" so that they might "focus attention on improving fuel economy." Companies could petition for a one-year suspension of the standards and, as Train acknowledged, plants could more easily convert to coal. The law did not, however, include the two provisions to which Train had most strenuously objected, the most pernicious proposals against the environment. Train's impact was obvious, although he did not enjoy such a public spat with many in his own party, later defensively noting that he had always acted in consultation with the Republican leadership on Capitol Hill. "It was a fairly unhappy time," he recalled.[35]

Reaction to the new legislation was swift. While Senator Muskie declared that the compromises inherent in the law "do not do the damage that some in the administration would have proposed," the Clean Air Coalition denounced Train's role. "With minor reservations, EPA today joined the rest of the industry-oriented Nixon administration," the environmental group declared.[36] In the end, such criticism was unduly harsh, although in the broader context of the energy situation Train might have accomplished more. Environmental restrictions did not cause the energy crisis but did make solutions more complex. In assessing Train's record, one should not forget the tremendous political pressure building against environmental protection. Nixon, already drifting to the political right before the oil embargo and Watergate coalesced, recognized the growing number of ideological conservatives as a defense against both. The White House had long before abandoned the environmental vote, making any sacrifice for the natural world a difficult sell. The auto standards raised legitimate questions of technological feasibility and, given the oil shortage, the additional use of coal was inevitable. Train was no scientist but he was smart enough to contract for studies of six major industries, including steel, oil, chemical, and electric power.

He may have lacked personal expertise but had solid data. To fault Train for recognizing the scientific and political reality was to expect too much.

Many conservatives certainly did not share the opinion of the Clean Air Coalition. "You idiot," wrote one. "Auto manufacturers do not need a push from Uncle Sam," wrote another. "I realize feeding at the public trough is an easy way to make a living but, if you'd get a job, a productive job, you could be an asset to your country."[37] Closer to home, in the White House mess, Senator Scott applied pressure more directly. Suddenly standing as a group of officials ate breakfast, Scott turned to Train. "Anyone who will not support the administration on this ought to quit, and I mean you," he declared. "They can find some ambassadorship for you, maybe Russia or China." Not surprisingly, Train did not return the angry fire. "Bill," he quietly replied, "you're way out of line."[38] Train, the calm and reasoned moderate, may have compromised but still won little praise from the environment's harshest critics.

The failure of the administration's proposal—and the new legislation—to address adequately the issue of consumption was hardly Train's fault. While Nixon and Congress did not make conservation a priority, Train consistently stressed the need for sacrifice. America held 6 percent of the world's population but constituted more than a third of its energy demand, Train noted in one editorial published in the *American Lung Association Bulletin*. "We will have to alter old habits," he wrote, "if we are to get to the root of the real problem."[39] It was a refrain repeated in speech after speech. Speaking at the University of the South just as Congress debated the administration's proposals, Train stressed the necessary costs. "These costs are not going to be borne by someone else," he stated, "but by ourselves."[40] Train planned to speak his mind even if much of America did not want to hear it.

If some environmentalists thought Train's compromises a capitulation to industry, they needed only note his actions on other energy-related issues to understand his commitment. In one example, as Congress debated amending the Clean Air Act, EPA ordered Chrysler Corporation to recall over 800,000 1973 model passenger cars, almost half its total output of Chryslers, Plymouths, and Dodges. A defect in the automobiles' pollution-control systems resulted in excessive nitrogen oxide emissions. A temperature-sensing device necessary to activate an exhaust-gas recirculation system failed to operate properly. Chrysler complained that it installed the device at EPA's behest, making the agency "partially responsible." To correct the problem would cost millions of dollars. Train bungled the announcement of his order, leaking it at a meeting of the National

Conference of Mayors. Bystanders heard him remark, "I just let it dribble out
. . . I blew it." It was a rare breach in Train's well-practiced political etiquette but
it did little to soften his defense against vigorous industry protests. When the
official announcement came, Train recovered his political wits but was equally
firm. Rather than increasing tensions, he claimed that the company had been
"cooperative in every way." At the same time his resolve was obvious in a letter
to Sidney Terry, Chrysler's vice president. "It is my conclusion," Train wrote,
"that sufficient data has been provided . . . to determine that within 50,000
miles virtually all vehicles equipped [with the defective device] will emit nitrogen oxide considerably above the Clean Air Act's standards."[41]

Regarding another controversial issue, many environmentalists complained
that Train too readily abandoned federally mandated transportation control
plans. According to the Clean Air Act, states were to provide such plans when
urban areas proved unwilling or unable to meet air quality standards by 1977. After Los Angeles failed to comply and a federal court ruled California's subsequent
plan inadequate, EPA assumed responsibility. When the Natural Resources Defense Council secured a court order forcing EPA to proceed with similar plans
for over thirty other cities, Train faced a dilemma. He had no choice but to de-
vise plans that included such provisions as changes in traffic patterns, increased
emissions inspections, parking restrictions, and other unpopular measures. At
the same time, plans for the most egregious violators promised significant economic repercussions. For Los Angeles to meet the standards, Train noted, the
plan "would have essentially shut down the city." It appeared a no-win situation.
The growing number of ideological conservatives cried foul, noting that the federal government now had a role in approving everything from shopping centers
to parking garages. Although Train had no choice, Republicans inserted language
into EPA's appropriation bill that essentially prohibited enforcement of any such
plan. This angered environmentalists, who protested that Train had too easily
folded in his environmental advocacy. Events had, arguably, overcome Train,
who was caught between judicial fiat and public opposition. In many respects
Train was a victim of circumstance, the criticism he endured unwarranted.[42]

Certainly Train deserves praise for advocating strip mining regulation even
as he compromised on the additional use of coal. Several years before, he had
pushed for a bill to regulate, not ban, the practice, a position that understandably
angered many environmentalists, given the momentum then for environmental
protection. Now, however, all recognized it as an improvident time to broach
any form of restrictive legislation, especially given the failure of Congress to act

previously. Train's renewed advocacy was courageous, and even environmental-ists applauded. When Democrat Morris Udall introduced a bill to create a new agency for strict regulation and enforcement, Train broke with Nixon, FEO, and most in his own party in supporting the proposal. Train demanded that the min-ing industry develop the technology to mine "deep coal," making strip mining less necessary. The White House had reason for anger. While mining affected pollution, the issue was more a matter for the Department of Interior than EPA. In addition, Train's renewed public opposition to his superiors, following so soon after the debate on the Clean Air Act, prompted reports of "Watergate anarchy." The debate persisted as the scandal engulfed the administration, with Train's position augmenting Nixon's public relations woes.[43]

To suggest that Train showed courage in standing up to Nixon or that envi-ronmental criticism too often failed to consider context is not to absolve Train completely. The energy situation and Watergate created a complex milieu of factors that presented both opportunities and restraints. Even as many of Train's actions on behalf of environmental protection deserve praise, in some instances he might have done more. He did not vigorously protest legislation to exempt the Alaskan pipeline from NEPA's impact statement process, which Congress passed and thus granted the project's final approval. Even as transportation plans truly posed untenable restrictions, innovative solutions such as surcharge taxes on commuter parking might have helped. Train, who had long advocated tax incentives to encourage environmental protection, proposed such an alterna-tive only to let it die on the vine as the number of conservative critics grew in Congress. "I would say that I was responsive to public opinion," he sheepishly admitted in one interview, "including congressional opinion." Among the con-gressional leaders was his nemesis Scott, who termed even the tax a "drastic measure." Unlike other issues, the White House exerted no pressure and the decision to retreat was Train's own. After meeting with several dozen California mayors, he declared that "the public simply wasn't ready for that kind of strong medicine."[44]

Most importantly, Train might have noted that at least some of the oil short-ages were artificial. He periodically blamed import quotas, low refining capacity, the international situation, and price regulations in addition to inflated con-sumption. He did not explain, however, that the "Seven Sisters," the world's seven leading oil companies, were an additional factor. With the influx of mi-nor independent competitors threatening their market share, the companies used the embargo to claim a shortage of oil. They stopped their practice of supplying off-brand independent dealers with their surpluses and simultaneously deactivated a large percentage of their brand stations leased to such dealers. In time, they planned to reintroduce their own self-service "independent" stations, which would sell gas for a few cents less than the major brand but not offer credit or extra service. For the present, however, this strategy meant less gas, higher prices, and longer lines. The scheme was worthy of criticism but, perhaps out of deference to his superiors and their close relationship with the culprits, Train largely remained quiet. When Democrats, citing new polls that questioned corporate innocence, introduced legislation to create a corporate "windfall profit tax" to recapture ill-gotten gains, Nixon swiftly vetoed it. Train might have voiced outrage but only a whimper came from EPA.[45]

In the end, perhaps Train's response to the energy crisis as Watergate unfolded was a story of the politically possible. Perhaps it was a testament to his brand of moderation and bipartisanship. From any perspective, however, Train played a pivotal role at a time when environmental advocates needed all the help they could get. "We are recognizant of your position," wrote the Nature Conservancy, "and we want you to know that we sincerely appreciate all that you are doing." According to the journal *American Forests*, "We can afford to lose a brick here and there providing the building still stands when the energy furor quiets down."[46]

Not every brick, of course, dealt with the energy crisis; the threat to the building of environmental protection continued to extend on numerous fronts. Even as energy concerns dominated his agenda, Train faced compromise and criticism on almost every one of his decisions. Facing stiff resistance in the House and Senate agriculture committees, he showed tremendous courage in banning the use of the chlorinated hydrocarbons aldrin and dieldrin. Farmers thought the pesticides, placed in the ground along with corn seeds, vital in the fight against termites and moths. The principal manufacturer, Shell Chemical Corporation, complained vigorously that the decision was "blatantly unfair." Protesters began disrupting Train's press conferences with shouts of "Liar!" Armed with the best data available, Train countered that the pesticides were carcinogens. A second Train decision to ban aerosol sprays employing vinyl chloride was less courageous but equally important. The manufacturers had already acknowledged the chemical as a carcinogen and begun a voluntary recall. According to Train, however, "the process was not moving fast enough."[47] Not all decisions, however, pleased environmentalists. When Train allowed emergency use of the banned pesticide DDT to control the tussock moth in the Pacific Northwest, environmentalists cried foul, some shedding tears at the announcement. In all prob-ability the threat from the moth had already diminished with the end of its natural growth cycle, the ban thus unnecessary. The decision was unfortunate but, nevertheless, understandable. Leading the push for action was Oregon Gov-ernor Tom McCall, a champion of land use planning and an environmentalist in his own right.[48]

Given the demise of land use legislation several years before, Train deserves considerable credit for trying to resurrect the issue in the middle of energy and political crisis. He created a new EPA division dedicated to land use problems and once again broke with the administration by supporting a new Morris Udall bill to encourage regulation. "The last ditch role is a familiar one for Train," sur-mised pundit Marquis Childs. "Since the onset of the energy crisis he has been fighting on one fall back position after another." Predictably, the proposal was no more successful than the earlier legislation, ensuring perhaps the greatest failure of the Nixon administration in environmental protection.[49] Undoubtedly search-ing for consolation, Train sought to use the defeat as fodder for raising EPA to cabinet-level status, a new agency to include CEQ and the responsibility for pol-icy formation. Ignoring that he had interjected himself into the land use debate, Train claimed that EPA "is being pushed, willy-nilly, into policy making—nota-bly in the field of land use—on a scale not envisioned when the agency was cre-ated in 1970." An obvious attempt to expand his own jurisdiction, the proposal left him open to criticism. "I'm not empire building," he claimed. "I don't think this idea has any practical likelihood of coming into being in the near term."[50]

Any thought of augmenting EPA's power was hopeless, the forces already aligned against the agency growing every day. Angered by Train's decision to ban the pesticides, conservative southerners in the House Agriculture Committee demanded that EPA comply with NEPA's impact statement requirements just like every other federal department and agency. It was an ironic twist, turning the law against the environment's champion. Train insisted that the law did not bind EPA but, faced with the possibility of retaliatory budget cuts, relented.[51] Train had to watch every move, the potential for criticism ever present. In one example, the *Washingtonian* published an article accusing his family of profiting from the energy crisis. It would have been an outrageous claim if not for docu-mentation that Aileen had purchased one hundred shares of Exxon Corporation stock. A reporter, it turned out, had rifled through Train's trash and recovered a notice of the transaction. Train had mistakenly received the notice intended for his financial advisors, who managed all the family's affairs in a blind trust.

Employing his connections as he had so often in the past, Train called several journalists and explained the situation, fortunately in the end diffusing any further embarrassment.[52]

His critics' numbers growing, Train's ability to maneuver in Washington's insular political world helped immensely. Recalling Train's environmental advocacy at "one of the most delicious and festive dinner parties I can remember attending," one administration official found himself persuaded. "You always do this so smoothly, so firmly and pleasantly, that I am proud to be with you."[53] Occasionally Train sent colleagues gifts of vegetables from his Eastern Shore garden. Aileen, for her part, delivered figs and tomatoes to key Washington allies and friends. In one instance, she hosted a "surprise dance" for her husband at the tony Chevy Chase Club. There the guests, including the usual assortment of Washington insiders, presented Train with a crown and robe and proclaimed him "King for an Evening." Throughout, Train never lost his common touch, whether joining with the community to dedicate a restored historic bridge in Harpers Ferry National Park or with Princeton students in the university's local "tap room." In the latter instance Train stayed late into the night. "Interesting for me and I think for them," he wrote in his diary.[54]

Aileen remained one of Train's biggest assets in winning over potential critics. She too had a common touch that belied her patrician background. Her commitment was equal to her husband's and was reflected most notably in her founding of Concern, Incorporated. Aileen had served on the boards of the Maryland chapter of the Nature Conservancy and the Chesapeake Bay Foundation and thus knew what it took to run a nonprofit, tax-exempt organization. She founded Concern to promote environment-friendly consumer practices and enlisted a number of prominent Washington women as volunteers. Concern published a regular newsletter, *Ecotip*, that described the best products to buy and eat. The organization produced a movie about the threat to drinking water and sold over 100,000 calendars highlighting its cause. Aileen met with Republican women's clubs reminding them that the GOP had played a prominent role in environmental protection and should not abandon the issue now. At her many dinner parties, in short, Train was not alone in his environmental advocacy.[55]

Weighing criticism, deciding when to press his case and when to retreat, was never easy for Train. In the international realm, at least, he continued to operate relatively unscathed. His success in implementing the 1972 American-Soviet environmental accord was impressive but only one example. Despite a President now consumed with scandal, dozens of exchanges of personnel and technology regularly took place under Train's watchful eye. The exchanges ran the gamut from the protection of swans to the "urban environment." Train regularly visited the Soviet Union and became friends with his Soviet counterpart, Yuri Izrael. Practicing the personal diplomacy for which Nixon was famous, Train entertained the Soviets on his farm—"without a KGB watchdog," he later recalled. He exchanged gifts much as he did with political colleagues and, when the occasion demanded, embraced Russian culture. "Some years ago Russ entertained an enthusiastic Russian who insisted that his host match him glass for glass of vodka," a friend remembered. Unaware that Train clandestinely poured his drink onto the carpet, the Russian was "awed by Russ' staying power." Full of admiration, "He staggered to his feet, close to oblivion, and embraced Russ, saying in a heavy Russian accent, 'Russ, I love you.'"[56]

Train's success with the Soviet environmental accord led to similar agreements with other nations. In the midst of the energy crisis, Train flew to Bonn to sign an agreement with the Federal Republic of Germany. A round of meetings and receptions followed, although perhaps lacking the degree of drinking diplomacy required for the Soviets. Only weeks later he received the Dutch Order of the Golden Ark, presented to him at the famous Soestdijk Palace. America's relationship with Japan continued to grow, meanwhile, Train once again employing his brand of personal diplomacy. "Mrs. Yamanaka and I are still yearning for your handmade cake and taste of tea," Defense Minister Sadanori Yamanaka wrote Train after one visit. Frequently traveling because of his CCMS duties, Train still enjoyed his international role. In this realm, at least, he faced little opposition and criticism.[57]

Train remained proud of his role in the 1972 Great Lakes Water Quality Agreement with Canada, declaring it "one of our major international agreements." He visited the area several times, once flying over its waterways for an inspection. In each instance he praised progress under the agreement. Progress was undoubtedly real, although Train might have stressed that, in regard to this particular international agreement, retreat was also evident. The administration, the Canadians complained, had fallen behind in its promised construction of waste treatment centers and had reduced its contribution to the International Joint Commission, necessitating cuts in its Great Lakes field office. It had also failed to deliver monitoring equipment required in the agreement. Without the requisite money, Train undoubtedly knew, compliance was impossible.[58]

In fact, Train remained quiet about one of the most significant aspects of Nixon's environmental retreat, money allocated in the Water Pollution Control

Act Amendments for waste treatment centers. After Congress overrode Nixon's veto two years before, Nixon ordered Ruckelshaus to spend only half of the $18 billion allotted, citing the phrase in the bill "not to exceed" as justification. This "impoundment" of half the money outraged environmentalists and local communities set to receive the funds. New York City sued and, just before Train arrived at EPA, the U.S. District Court for the District of Columbia ruled in its favor. The administration appealed and the litigation continued until 1975, when the U.S. Supreme Court agreed that the impoundment was illegal. Rather than noting that urban growth continued to outstrip construction of waste treatment centers and that the need for the money was great, Train defended Nixon. "The president," he explained in one interview, "has to put waste treatment needs into the overall budget priorities that he must deal with." The money the White House did spend, he continued, was more than the federal government had ever before spent. When pressed, Train acknowledged that while he had sent a letter to Nixon advocating spending the full $18 billion, he had not attempted to meet with him or press the matter further. This was odd given the importance of the issue, which included a major question of constitutional authority, and the manner in which Train had so publicly broken with the administration over air pollution. Some question existed over whether local communities "could manage and handle the money as a practical matter," Train insisted. "It might have been impossible at such high levels."[59] To the degree that this was true—Train conveniently ignored—much blame fell on EPA. Several states complained that EPA's extensive bureaucracy and paperwork slowed their own clean water programs, a fact documented in congressional hearings. Years later, when asked about the issue, Train was clearly defensive. "I don't recall," he curtly replied. This in itself was somewhat disingenuous. The name of the long-running case was, after all, *Train v. City of New York.*[60]

Despite his many achievements, could Train have accomplished more? The answer invariably turns to Watergate, which by the summer of 1974 was building to its inevitable climax. With a new special prosecutor joining the House Judiciary Committee in subpoenaing his White House tapes, Nixon tried releasing transcripts. Eliminating particularly embarrassing passages such as profanity or insults, he hoped to comply without further weakening his position. Ensuring accuracy without revealing evidence of a crime was difficult, the result no smoking gun but hardly a favorable portrait of the President. Out for public consumption was the vindictiveness, anger, and political intrigue that Train had encountered over the previous months. When the transcripts did not calm the furor, Nixon claimed executive privilege in refusing to release the full tapes. In late July the news arrived. The Supreme Court had ruled against Nixon; the tapes were public property.[61]

Nixon knew that the end was at hand. The tapes contained irrefutable evidence that he had attempted to cover up White House connections to the break-in at the Democratic National Committee. The charges were serious—obstruction of justice and abuse of power—and impeachment was imminent. He had no choice. He would have to resign.[62]

The drama played out in newspapers and televisions across the land. Train, however, had a front-row seat to history. He knew more than most that Nixon, in large part, directed his anger at liberals, whom he blamed for his predicament. In a sense it made his conservative inclinations more firm—and the task for Train more difficult. Nixon wrote Train that it was "particularly important" for EPA to contain costs.[63] Only weeks before his resignation, Nixon met with OMB officials and instructed them to make further cuts in the agency's budget. When Congress passed additional funds, he vetoed the budget bill, one of his last official acts. "What a bargain!" one pundit sarcastically wrote. "For the price of sacrificing American land and its inhabitants, we increase the GNP." Nixon, packing his bags, little cared. Perhaps, in his mind, it was payback time.[64]

In one sense, "impeachment politics," as Washington Senator Henry Jackson termed it, limited Train's autonomy. As the scandal slowly eroded the administration's support over a matter of months, Nixon readily tossed aside any vestiges of environmentalism as a way to garner congressional support against his impeachment. The scandal, the *Economist* surmised, "has forced [Nixon] to show more consideration than he might normally show for the interests of right-wing senators, both Republican and Democratic, whose votes he may desperately need before the year is out." It transferred power to Congress, "which by its nature is even more open to [industry] pressure than the executive branch."[65] In another respect, Watergate produced a "vacuum of leadership," in the words of Norfolk, Virginia, mayor Roy Martin, speaking on behalf of the National Conference of Mayors. Nixon did not have the time or energy to devote to any domestic matters, Martin believed, and thus the environment suffered.[66]

Train, however, who better understood Nixon's antipathy toward environmentalists, believed that a vacuum did indeed exist but that the result was positive, not negative. At one point, somewhat facetiously, he even declared Watergate "the best thing that ever happened to EPA." The White House's preoccupation allowed Train to disagree publicly with the administration whereas before Nixon would not have tolerated it. Train was popular and the President could not afford to terminate him without engendering further criticism. The year before, Train had pressed Vice President Agnew for more autonomy. Now Train told the New England Energy Policy Council, "There is more responsibility and flexibility left to agency heads and that is all to the good."[67] Had Train pushed Nixon on water pollution as he had on air pollution, it is difficult to predict the White House's reaction. Had there not been Watergate, however, Train's days in the administration undoubtedly would have ended.

One fact is clear. Train's independence not only protected the environment but also saved his career. For years he had worked to be part of a team. Just when the team faced defeat, however, the public perceived him to be separate. "Train survived because Richard Nixon did not," wrote one colleague. "What had been to Ruckelshaus a voice of encouragement was to Train a lifeline."[68]

Train heard the news of Nixon's decision to resign while vacationing at the Ausable Club, high in the Adirondack Mountains not far from where he had spent his summers as a boy. He immediately called a colleague at the White House, Bryce Harlow, and asked if he should return. Harlow replied that he could think of no reason but "if I was not there and needed . . . I would always regret it." Gathering with his friends in the main room of the club, Train watched Nixon's announcement on television. He was "sad" while amazed at Nixon's "real gap in judgment." The scandal was, he figured, "an outgrowth of Nixon's fascination with the exercise of power." After the announcement, Train immediately drove to Albany and spent the night in an airport hotel. The next morning a flight brought him to Washington and his office, where he watched on television Vice President Gerald Ford sworn in as the nation's thirty-eighth chief executive. After a brief meeting with top aides, he returned to the airport and the Adirondacks. No one needed him; in fact, the transition of power was surprisingly smooth.[69]

Back in the mountains, Train had time to reflect. A sense of pride in the strength of American democracy and relief at the ordeal's final end brought a new perspective. He owed a lot to Richard Nixon—as did the nation's environment—but he must have wondered if, now, being both a Republican and environmentalist would be simpler. Once again the future was uncertain but hopefully better than the frustrations of the past.

8

A New President, a New Right

Gerald Ford and Russell Train appeared to be cut from the same cloth. While the new President lacked Train's pedigree, he had earned his status as a Washington insider. Colleagues from both sides of the aisle liked Ford's genial manner. Democratic Speaker of the House Carl Albert was a close personal friend, declaring Ford, "a very fine man to work with."[1] With a quarter-century in the House of Representatives, Ford was not the overt, aggressive partisan like Nixon. Rugged, even tough looking in appearance, Ford revealed a much more gentle nature than his background as a lineman on the University of Michigan football team suggested. Like Train, he enjoyed the cocktail circuit and Washington's social scene, his standard martini with lunch a concern for image-conscious staffers. "He doesn't twist arms," concluded Illinois congressman Edward Derwinski. He was an "open tactician" but more subtle and clever. In his vice presidential Senate confirmation hearings only months before, Ford appealed to the Democratic Congress. "There has to be a two-way street," he testified, promising to seek the advice of legislators. Train could not have said it better; both men were apparent masters in the traditional ways of Washington.[2]

No one doubted Ford's conservatism but neither had Ford joined in Nixon's recent denunciation of environmentalists. "America must change its way of living or smother in its own wastes," he had stated on the first Earth Day. Ford had supported Nixon's early environmental agenda and now declared himself a "moderate on domestic issues." Although relatively quiet on environmental matters, he had voted to override Nixon's veto of the Clean Water Act two years before. As Train battled Nixon over weakening the Clean Air Act, Ford had surprised many with words of support. "We must establish stringent air quality standards, particularly for pollutants that are hazardous to health," he declared as the debate raged. As if to highlight his moderation, Ford soon selected New York governor Nelson Rockefeller as his Vice President. Long recognized as a pragmatist and not an ideologue, Rockefeller shared Train's patrician background as well as his overall political philosophy. Train welcomed the selection.[3]

Train did not know Ford well; the new President had never served on any of the environment-related committees in Congress. Nevertheless, the two had met several times and Train, while CEQ chair, had periodically telephoned Ford, then Minority Leader in the House of Representatives. Wasting no time in gaining Ford's favor, Train quickly offered congratulations. "Your statement today was right in every respect and exactly what the country needed to hear," Train wrote after Ford's inaugural address. "You have my full support and I look forward to an opportunity to meet with you at your convenience."[4] Working indirectly, Train called his friend Rogers Morton, who served on Ford's transition team. Morton recommended Train put his thoughts into writing. Train then talked with Donald Rumsfeld, the designated chair. Rumsfeld likewise encouraged his old acquaintance while adding, "I wouldn't touch the top White House staff job with a ten foot pole."[5]

Train's memo on "how to strengthen and make more effective the relationship of the President to agencies and to Congress" reflected his frustrations with the Nixon White House. Agency heads deserved "open and direct communication on a regular basis" and "maximum freedom in decision-making consonant with overall Presidential policy." The President should allow "full, advance consultation with Congress, on as bipartisan basis as possible." Finally, on a point that Train acknowledged had "an element of self-interest," he asked to "attend Cabinet meetings as a matter of course."[6]

Once again Train's luck had turned; Ford shared most of his concerns. Ford sought advisors who "had no qualms about telling me when I was wrong." There needed to be "openness and the free movement of people and ideas," their differences a source of strength. "What I wanted in my Cabinet were strong managers who would control the career bureaucrats and not become their captives, people who knew how to build support in Congress and the media," he later recalled. "I would leave the details of the administration to them and concentrate on determining national priorities and directions myself."[7] Train had a hint of this attitude early. After meeting with agency heads in the White House's Blue Room, Ford agreed to photographs with Train and his staff. When Ford asked where he should stand, Train replied, "Right in the middle, Mr. President." Ford's retort was telling: "That's where I am used to being." Within days the media reported on the President's open style. "He makes dozens of telephone calls to expose himself to many points of view and holds frequent meetings with his ·advisers," reported *Time* magazine.[8]

When Train learned that John Sawhill of the Federal Energy Administration (FEA) had already visited the Oval Office, he decided it best "not to just wait on an invitation but try to generate one." Employing his connections once again, Train called Dave Parker, Ford's scheduler and a former colleague at the Interior Department. With the appointment set, Train confirmed the meeting to the *Washington Post*. This annoyed Ford, who commented to staffers that "leaks are one thing but floods another." Train later denied the initial leak but the story clearly noted the importance of the environmental threat the nation—and Train—faced.[9] Once in the meeting, Train was characteristically diplomatic. After apologizing for the *Post* story, he mentioned the Great Lakes as an area of environmental progress, an obvious attempt to pique the interest of the Michigan-bred President. Wisely avoiding specific issues, he thought it "important to establish a rapport." Appealing to Ford's conservatism in another attempt at ingratiation, Train proclaimed himself "no emotional crusader," recounting his Republican and professional credentials. The public, he continued, perceived a Republican promise of balance in environmental policy as an industrial bias. "I suggest that when you are addressing environmental matters, you take a straight pro-environment line and have me do the balancing." Explaining the ways of Washington to a sitting President, a man with more political experience, might have seemed presumptuous. If Ford thought so, he did not show it, quietly puffing on his pipe.[10]

Train pressed ahead, noting the need for bipartisanship and his desire to attend cabinet meetings. In a clear attempt to gain a ride on Air Force One, he mentioned his plans to attend the same mass-transit conference in Pittsburgh that the President was to attend. Finally, he asked for written confirmation that he would remain in charge of international environmental affairs, a tactic he had employed with Nixon upon arriving at EPA. Ford's response to all this was a pleasant surprise. An invitation to cabinet meetings and Air Force One soon followed, and Train left the room assured that his diplomatic role remained. The "high point" for Train came near the end of the meeting. After Train commented that he did not know when the two would meet again, Ford turned to his assistant. "I want to see Russ on a regularly scheduled basis such as once a month or at least every two months," he ordered. Turning back to Train, "If you ever have any issue you think you should discuss with me, all you have to do is call up and say so." Train left the Oval Office thrilled, happily surprised when he realized that energy and budget officials had been waiting for his lengthy meeting to conclude.[11]

It did not take long, however, for Ford's promises and Train's renewed clout to face a test. In early October, a colleague called Train and asked if he had read the *Wall Street Journal*'s article on Ford's impending economic message to Congress. Train had just attended an economic conference at the Washington Hilton with several of Ford's key staff and no one had said anything. He assumed the statement innocuous. The *Journal*, however, told a different story. Ford, it read, planned a "warmed over version" of the Nixon administration's proposals to amend the Clean Air Act. Worse, the message would call for the permanent use of tall stacks, the very proposal that Train had so strenuously—and successfully—resisted the previous year.

Flabbergasted, Train called OMB, Rumsfeld, and Ford's top energy advisor, Bill Simon. Despite his earlier comments, Rumsfeld had just accepted the position as Ford's Chief of Staff. "Rummy," as Train now knew him, and the others professed to know nothing. When Train asked for a copy of the statement, however, Simon became evasive. It was, Train complained, "a helluve (*sic*) note for a major Presidential policy statement directly involving EPA's specific interests to be prepared without any consultation with EPA whatsoever." It appeared a direct contradiction of Ford's promises. A series of phone calls commenced and continued throughout the afternoon before Train finally got one official to read the entire text. "I am going to scream like hell in a few moments," Train remarked to his staff. When the official came to the tall stacks language, Train "went into my act." In a high-pitched scream, he belted out, "Wh-a-a-t? I can't believe it!"[12]

The outburst had its intended effect. At seven that night, Train met with Ford's economic team in the White House's Roosevelt Room. When Train asked to see a copy of the final draft, Simon suggested that he just look at the "fact sheet." Train persisted until they agreed to go over the draft line by line. "I'll be darned," one official declared, "it does talk about tall stacks." Fuming but keeping his calm, Train noted the health implications just as he had with Nixon. "No one there," he later wrote, "had any familiarity with the matter other than Simon." In the end, Train once again prevailed. The message did not reference tall stacks, a significant victory for the environment.[13]

Not all environmentalists appreciated Train's efforts. "We were hoping for a really bold move by Ford," commented David Brower, who had obviously read the *Journal* article. "What we got was warmed-over Nixon."[14] The final draft included some language that Train did not appreciate, but he had served notice that the White House could not ignore him. It was a victory in more ways than one; he had sent his own message.

As this incident resolved, Train attended his first cabinet meeting. He was the only agency head present, an obvious indication of his stature, if not Ford's promises. It was, Train later recalled, "positive reassurance as to my own role." Ford noted his intention to require all agencies to develop "inflation impact analyses" for every proposal, an obvious reaction to the rising cost of living and its implications for the economy. Train correctly perceived that energy officials might try to use such studies as a way of "rolling back" environmental regulations. He did not believe this a great threat, however, even viewing the analyses as a "two-edged sword." The studies might prove a "useful additional weapon against various programs, such as highways and dams, which can have seriously adverse environmental effects."[15]

Within months, Ford announced plans to reorganize the White House's Domestic Council, including giving the new Vice President Rockefeller "a major role on domestic affairs." Rockefeller followed by promising "meetings with substantive groups on the Hill." Train found Rockefeller's habit of stirring his coffee with his glasses odd but welcomed his apparent openness. Rumsfeld echoed this sentiment, promising, "The President does not want the White House staff interfering with agency operations." Yes, it appeared, Ford's comments were more than just rhetoric.[16]

Train was not shy about letting the world know. "[Ford's] record in Congress on the environment is conservative," he acknowledged in one interview. "But people's perspectives change when they move into the White House." Acknowledging that few substantive issues had yet risen, Train praised his "very good relations" with Ford. "I'm in the middle of everything." In speeches and interviews, Train stressed that Ford was no Nixon. "The significant thing is that [Ford] has said he wants to see me once or twice a month and to call him when certain issues become hot," he declared. Ford had better information, with EPA "much more part of the decision-making now."[17] A new President and a new attitude, Train implied, meant a new direction for environmental policy.

If the new President provided hope, Train was under no illusion that any shift in policy would be easy. Even as he praised Ford, he harkened back to the task that remained. Environmentalism, he told EPA employees just after Nixon's resignation, had "survived its first onslaught in good shape." Nevertheless, a new President did not mean a "free ride."[18] Indeed, as Train worried, the economic and political clouds continued to gather, a storm that even the most sympathetic President could not prevent. Inflation now stood at an annual rate of more than 12 percent. Lost in the news of Nixon's resignation was the announcement of July's wholesale price index—a whopping 3.7 percent, the second biggest monthly jump since 1946. The stock market was jittery at best, the Dow recently plummeting below 780. The energy situation had improved somewhat since the end of the Arab oil embargo but demand continued to grow despite recessionary indicators. Gas remained expensive, with the nation paying almost $30 billion for foreign oil in 1974, ten times the sum spent only five years before. Worse, unemployment showed ominous signs of rising. One trade association representative described Ford's situation succinctly. "He's on a bus full of people that's flying down the road pretty fast," he said. "It's got some mechanical problems and, all of a sudden, the driver jumps out. Ford's got to get in the seat and keep driving."[19]

In recent years historians have noted that the economic stagnation that refused to ebb challenged the very core assumptions of environmentalism. According to Hal Rothman, environmentalism spoke for the spiritual and non-utilitarian values of natural resources, contrary to historic patterns of economic behavior. In times of prosperity, such as the quarter of a century following World War II, the nation could easily cherish these values as signs of a compassionate and rational culture that understood its technical limitations. In times of economic strain, however, they became a luxury the nation could no longer afford. The goals of environmentalism increasingly appeared class based and insensitive to more basic concerns of economic survival, a difficult obstacle for environmentalists to overcome.[20] Nixon had left but the economic problems remained. The environmental retreat he helped launch late in his term involved more than the temporary policies of one administration—an ominous cloud for environmentalists indeed.

In some respects, the accomplishments of the Nixon era diminished the sense of crisis that had surrounded the growth of environmentalism, turning much of the debate off the front page and into the legal arena. Lawyers with briefcases had replaced protesters with signs. Train knew the importance of legal implementation; it had been the reason for his move to EPA. Maintaining public enthusiasm was no longer as easy, however, with the media recognizing other headlines as more gripping. In essence, a more developed environmental movement now stood as one of many special interest lobbies, as it struggled to maintain momentum regardless of the chief executive.

At the same time, the economic malaise fueled the growth of the more stringent strain of Republican conservatism, the cause—which many had assumed moribund—of senators Hugh Scott, Barry Goldwater, and others. "Economic discontent helped the conservative movement," concluded historian Michael Schaller, "which grew in organizational strength and popular appeal during the 1970s." For most of the postwar era, big business had cooperated with big government and big labor to promote economic growth. Until the energy crisis, this "growth coalition" produced high corporate profits, rising wages and consumption, increased government spending and investment, and, initially, an embrace of environmentalism. With such profits, most business leaders felt little urgency to ally with such unpopular conservatives as the Arizona Senator. With the end of the boom, however, corporate America began to transform its economic and political strategies. Executives worried by lower profits viewed themselves as being under attack from global competition, labor militants, consumer advocates, and, in time, environmentalists. They reacted by abandoning the "growth coalition."[21]

Reflecting this change, big business overcame the narrow interests of individual industries and began to create umbrella lobbying organizations, the Business Roundtable only one example. The number of individual businesses with registered lobbyists skyrocketed. New industry-supported political action committees grew, as did the number of nonprofit foundations and conservative think tanks. The American Enterprise Institute (AEI), the Heritage Foundation, and the Hoover Institution, among others, all preached the new gospel of supply-side economics, which stressed cutting taxes and spending. In the words of one journalist, the "AEI's deregulation-mad economists had been regarded as irresponsible anarchists." By the Ford administration, however, their annual fund-raising approached $10 million. Within the Republican Party a new conservative corporate culture grew, philosophically opposed to government regulation and, therefore, less open to compromise.[22] It did not bode well for the environment or Train. According to historian H. W. Brands, Train's own EPA was fast becoming the "bete noire of conservatives."[23]

Ford was hardly immune to this movement. If nothing else, his political genesis in Michigan's automotive-based economy made him sensitive to corporate concerns. He had, for example, voted against mass-transit funding, an action Train most likely hoped was an aberration. More importantly, Ford's selection of Rockefeller enraged and energized the new conservatives, the frustrations they harbored over Nixon's early domestic moderation now directly and forcibly aimed at the new President. According to conservative activist Richard Viguerie,

Rockefeller represented the "high-flying, wild-spending leader of the Eastern Liberal Establishment." Viguerie would not have been more upset "if Ford had selected Teddy Kennedy."[24] "[Rockefeller's] name was anathema to conservatives," Ford acknowledged in an understatement. Fourteen years earlier, Goldwater had urged conservatives to "take this party back." Now he warned that Rockefeller "would provoke a split in the party."[25]

Paralleling the new fiscal conservatives were the new cultural conservatives. Although this constituency remained in its political infancy during the Ford years, largely unaligned with either party, the foundations of a new Republican bloc were evident, if not in Nixon's Southern Strategy, then in events of Ford's day. Revulsed by Supreme Court decisions outlawing teacher-led prayer in schools and legalizing abortion, the latter only a year before Ford's Presidency, this group bemoaned the social libertarianism of the 1960s. In particular, they found First Lady Betty Ford's outspoken support for abortion rights reprehensible. Stronger in the South and the mountain states of the West, they focused on emotional cultural issues, in the words of Reverend Tim LaHaye, "easy divorce, abortion-on-demand, gay rights, militant feminism, unisex facilities, and leniency toward pornography, prostitution, and crime."[26] Strongest in the same areas where environmental concerns paled, where agricultural and extractive industries dominated, they seemed little concerned with environmentalism, often linking it with the "lifestyle liberals" they so abhorred. In fact, environmentalism drew its strongest support from the upper-middle classes—more prominent in the suburban and urban areas of the Northeast and along both coasts—who tended to accept civil rights, the counterculture, and the antiwar movement. Activists such as Howard Phillips, then chairman of the Conservative Caucus, had already begun efforts to unite the new fiscal and social conservatives, a movement that had the potential to seize control of the Republican Party from the "country club Republicans."[27]

Train may not have realized it but a political tsunami brewed. On the surface, at least, the Republican Party appeared in shambles, Watergate its apparent death knell. In the 1974 off-year election, Democrats made significant strides in Congress. A Gallup poll that year found that only 24 percent of voters described themselves as Republicans. Below the surface, however, economic and political forces promised a new dawn—albeit for a new Republican Party. Cracks had begun in the Democratic coalition that Franklin Roosevelt had assembled decades before, the growing influence of the New Left—the "lifestyle liberals"— annoying many traditional party loyalists. Even as only a quarter of Americans claimed to be Republicans, the number who self-identified as conservatives rose to almost 40 percent, higher than anytime since Gallup first asked the question in 1936. "If you are hungry and out of work, eat an environmentalist," read one bumper sticker, hardly a comforting sign to environmentally conscious moderates of both parties. Clearly a new conservative movement had momentum, drawing from sources that it never before had. Train's future, as well as the future of the cause he championed and the political party he called home, lay in the balance.[28]

Energy policy remained a primary battlefield of the political wars, of course, frequently driving a wedge between moderates and conservatives on both sides of the aisle. In this area more than most, Ford's promises of openness and moderation faced their greatest challenge. As the months passed, the same issues that had foiled Train during the Nixon retreat returned, proving without a doubt the potency of the new conservatives. Too often Train still found himself on the defensive, fighting to maintain past accomplishments rather than advancing his cause. His efforts hardly won praise from all Republicans. According to North Carolina Senator Jesse Helms, Train was the "Grand High Lama."[29]

Even as Ford continued to profess moderation, a tack rightward was evident. Although the recently passed Energy Supply and Environmental Coordination Act had not included the permanent use of tall smokestacks, many of Ford's advisors still pushed the issue. Characteristically, Train sought compromise. Working with the Energy Resources Council, Train won a commitment to scrubber technology but allowed a decade, until 1985, for company compliance. Surprisingly, Train found an ally in FEA's John Sawhill, who joined him in pushing for more forceful energy conservation. Mandatory controls, Train and Sawhill agreed, might be necessary. It was a strong stand—and it cost Sawhill his job. Conservatives on the White House staff forced his resignation, ensuring their dominance of administration energy policy. Newspapers even began speculating on Train's future, most agreeing with the *San Jose Mercury* that "so far [Train's] job seems secure." For the first time Ford exhibited a different demeanor in cabinet meetings. "No one in the Cabinet is to say we are willing to compromise," Ford declared in regard to his overall energy program. "If there are compromises, I will make them."[30]

Sawhill's political demise did not deter Train, who still enjoyed relatively easy access to the President. At Train's request, Ford agreed to an Oval Office meeting as the end of the year approached. Train reiterated the importance of environmental protection but Ford "didn't really respond." Train gave him the benefit of the doubt. "I had a feeling that he was fatigued, as well he might have been, and that I was covering so many different points that perhaps I was not getting his attention sufficiently to any one of them." Train explained that a number of auto companies had once again raised the issue of extending the Clean Air Act's emissions standards, a battle renewed from the Nixon years. Now they had requested that Train grant a five-year extension, a request that Train sensed Ford favored. Rather than push the issue, Train wisely deferred it. He successfully urged Ford to remain uncommitted until completion of EPA public hearings. Emboldened, Train suggested combining EPA with the National Oceanographic and Atmospheric Administration and "perhaps some other units such as the Fish and Wildlife Service" into a Department of the Environment. Given ongoing proposals for a Department of Energy and Natural Resources, an environmental department would fit with "your own announced desire to maintain balance." Train undoubtedly saw himself at the helm of such a department but Ford once again remained uncommitted. The meeting ended with a discussion of conservative discontent in Congress, Train acknowledging that Republican Bob Michel, a "very popular conservative," felt frustrated and sought better access to the White House.[31]

Several weeks after this meeting and only days after Christmas 1974, Ford called his energy and economic advisors to Vail, Colorado, where he and Betty were skiing. After a four-hour flight from Andrews Air Force base and a short but beautiful helicopter ride over the snow-covered mountains, Train arrived at the rented lodge. There he enjoyed a frank and open discussion, the group sitting with Ford before a large fireplace. After advisor Roy Ash stressed conservative market principles, a participant jokingly described him as "our free market enterpriser." Recognizing that he was opposed to Ash—and the majority—on the matter, Train objected without being objectionable. "I guess I am speaking as the house radical," he quipped to general laughter.[32]

If the nature of the meeting contrasted starkly with those of the Nixon years, the challenge for Train remained the same. Implementation of the Clean Air Act played prominently, initially more in regard to sulfur oxide pollution from stationary sources than auto emissions, Train's promised EPA emissions hearings then commencing. When FEA officials urged federal preemption over— frequently stricter—state implementation plans, Ford agreed with Train and denied the request. At the same time, however, Ford insisted that EPA hold formal hearings before determining the public health risk from sulfates in specific coal conversions, hearings that Train had hoped to avoid. That night, in the classic social climate in which he thrived, Train initiated a conversation with Milton Friedman, then one of Ford's speechwriters. Environmentalism, Train tried to explain, "was basically a very conservative concept." Friedman was receptive, although undoubtedly the others gathered around the twenty-three-foot Christmas tree would have taken exception.[33]

By early 1975, Ford had announced his intention to submit new amendments to the Clean Air Act reflecting the tall stacks compromise Train had negotiated. EPA had concluded its auto emissions hearings and Train prepared to announce his decision on the requested extensions. The decision was, naturally, another compromise. Explaining that the hearings had yet to establish that the catalytic converter posed its own significant health risk, Train repeated his conclusion of the previous year that the technology was important to the legislation's mandate. At the same time, however, he granted another one-year suspension, this time for the 1977 emissions standards, and recommended a program to reduce hydrocarbon, carbon monoxide, and sulfate emissions from the 1977–1982 model years. Compromising again was undoubtedly difficult but, in many respects, Train had little choice. States were already behind in finalizing their individual compliance programs according to the law, with only 91 of 247 air quality regions successful. This compliance rate not only evidenced conservative resistance but genuine difficulty in meeting congressionally mandated scientific standards. Train might have held firm but, in all fairness, it was a complicated decision that ran sixty-two pages.[34]

Ford, to his credit, left the decision to Train. The White House had, however, recently announced a goal of a 40 percent improvement in automotive fuel economy, a goal that many scientists claimed strict auto emissions standards made more difficult. Further complicating matters was an EPA regulation issued the previous year. The order required states to review in advance any plans for large construction projects that might hinder Clean Air Act compliance. Such "indirect sources" included new shopping centers, highways, or other facilities that might attract heavy motor vehicle use. Even with his new emissions compromise, Train felt it necessary to delay by six months this requirement. He knew that reaction from environmentalists would be "schizophrenic" but, left to play the role of Solomon, he genuinely had little choice.[35]

Train's compromise did not please everyone. Because leaded gas rendered catalytic converters ineffective, Train issued new regulations that mandated a fuel tank design for vehicles with the technology. No longer would the nozzle of a leaded fuel pump fit. Soon new EPA regulations encouraged the elimination of leaded gasoline entirely, because the lead itself proved a health hazard. This was all too much for many conservatives, who complained of increasing federal interference into their daily lives. Train felt the wrath personally. In the predawn hours one winter night, a phone call woke him from a sound sleep. A driver in Minnesota angrily explained that he had run out of gas, trudged through the snow and darkness to retrieve a gas tank, and returned to his car—only to find that the nozzle would not fit. "All because of your goddamned stupid EPA regulations," he yelled.[36]

More significant criticism came from Barry Goldwater, who termed Train's decision on the catalytic converter a "bureaucratic monstrosity." Reflecting the antipathy many conservatives felt toward federal regulations, Goldwater bemoaned in his regular newspaper column the "made-in-Washington blueprints." The "regulators" did not admit their mistakes because "bureaucrats never admit costly errors." Nevertheless, Goldwater wrote, Train's decision to extend the deadlines proved the fallacy of their ways. Not about to let such criticism pass, Train quickly replied in a "Dear Barry" letter. Congress set the standards, Train noted, and EPA "did not force industry to use catalysts." The air pollution problem was "serious" and "EPA and other federal agencies must share in the solution."[37]

Perhaps it was inevitable that the exchange was not over for the two Washington insiders. On a flight to his Eastern Shore farm, Train found himself next to the conservative hero. The conversation was surprisingly amicable, hostility perhaps easier to muster in writing than face to face. Goldwater acknowledged that Train "raised a number of technical points which have some basis in fact." Nevertheless, his defense of EPA hardly "exonerated" Washington's heavy-handed regulatory establishment. Goldwater informed Train that his busy campaign schedule prevented him from continuing his column, "so you don't have to worry about any trouble coming from me in the future." As Train answered respectfully, both undoubtedly knew that while the Senator might no longer take the lead in such criticism, many other conservatives were ready to jump to the fore.[38]

Ford certainly felt the criticism. In one instance, Train joined the President on Air Force One flying to Cincinnati to dedicate EPA's new $28 million National Center for Environmental Research. Afterwards Ford met with representatives of several environmental organizations, further indication that he was no Nixon. Everything appeared a success until Ford attended a White House Conference on Economic and Domestic Policy scheduled later in the day. After all the positive comments, Ford made a statement that won considerable press. "I pursue the goal of clean air and pure water but I must also pursue the objective of maximum jobs and continued economic progress," he declared. "Unemployment is as real and as sickening a blight as any pollutant that threatens this nation." The press noted the apparent contradictions in Ford's environmental and economic remarks, the speech muting the goodwill he had earlier earned.[39]

Ford's comments, Train later wrote, "did not seem unreasonable." In fact, however, they reflected a key component of the conservative critique. The speech seemed to acknowledge that environmental protection and the strong enforcement of federal regulations were mutually exclusive of economic growth. You could not have one with the other. Train may have been willing to give Ford the benefit of the doubt but Ford's comments represented the very antithesis of what Train had long preached. Only weeks after the Cincinnati trip, Train gave an interview to the *Washington Star*'s Roberta Hornig. "First," Train declared, "let me say that I feel very sincerely that there is no necessary conflict between the environment and energy goals." In the short term, before proper policy developed clean energy sources, compromises may be necessary. However, in the long term, "environmental concerns and energy concerns really complement each other." Conservatives such as Goldwater had never openly declared opposition to environmental protection but neither had they embraced the idea that federal regulations might encourage alternative industries.[40]

The industry lobbyists roaming the halls of Congress certainly did not welcome new competition; their millions in contributions funded conservatives of both parties. Environmental regulations, they cried, raised costs. It was a smart tactic that resonated in an inflationary economy and with Ford's own administration pressing the slogan "WIN—Whip Inflation Now." Increasingly Ford reflected reservations about the environmental price tag and stressed his mandated "inflation impact statements." Train may have hoped such statements a "two-edged sword" but now found himself more on the defensive than he had anticipated. By the end of 1975, EPA had missed the deadlines the White House had set for the development of statement criteria. EPA spokesmen sheepishly declared the agency "no further behind than anybody else."[41]

If the nation's approaching bicentennial promised to unite Americans, the Clean Air Act continued to divide them. The courts had ruled the previous year that the law did not allow "significant deterioration" of air quality in regions above federal standards, a decision that conservatives had unsuccessfully attacked. Now, when Train and EPA prepared regulations to implement the court's decision, new protests arose. Utah Republican Frank Moss introduced an amendment to postpone any EPA regulations. Senate Public Works Committee chairman Jennings Randolph answered with his own amendments to codify the court's decision while mitigating any negative economic impact. Train and the majority of moderates strongly supported the latter, of course, only to learn that Ford had joined with the conservatives behind the former.

The pressure on Train was immense. Dick Cheney, a leading conservative from Wyoming, had replaced Rumsfeld as Ford's Chief of Staff and symbolized the conservatives' new clout on energy issues. Told that his position undercut the President, even Train's well-practiced reserve faded: "That's a lot of bullshit!" Facing complaints that an EPA analysis supported the Jennings bill and not the Moss amendment, Train shot back, "I have been through one period of paranoia in this house and I don't want to do it again." The President, came the reply, "has given lots of leeway to his agencies . . . but now there is considerable feeling that the agencies have taken advantage of him."[42] Pressing Cheney while discussing the issue with moderate senators such as Tennessee's Howard Baker, Train arranged a meeting with Ford to discuss the issue. By this time, however, the White House had already released a letter supporting the Moss amendment. Train, Jennings, Baker, and the other moderates had little room to maneuver. "Where does that leave us?" asked Baker as he departed the White House later that day.[43]

Train was not ready to relent. Speaking with Ford separately, he agreed to support further changes to the committee's bill provided that he could publicly oppose the Moss amendment. In reality, the new compromise did little more than allow Train to save face. Ford remained on record supporting the Moss amendment; Train's efforts to change White House policy had failed. In the end, the moderates had more success in Congress than Train did with Ford. No Clean Air Act amendments emerged; the court's prohibition on "significant deterioration" was intact. In this sense, environmental concerns had prevailed. From Train's perspective, however, Ford, like Nixon, continued to drift to the political right.[44]

"It seems to me," Train explained to one White House advisor weeks later, "that there is too much of a tendency to equate conservatism with being protective of business."[45] When it came to the Clean Air Act, Ford's only concern appeared to be the auto industry, a direct influence of the conservatives on his staff. Even as the debate over "significant deterioration" raged, Train insisted that EPA had the right to test emissions standards on the automotive assembly line, not simply on various design prototypes. With only the latter, Train correctly argued, industry could easily circumvent the law. Pressing the matter,

Train submitted proposals for "selective enforcement audits" (SEAs). The White House, however, sat on the proposals, refusing to act as the months passed. By the summer of 1976, Train's allies on the Public Works Committee were as frustrated as the EPA administrator and they proposed legislation to ensure the SEAs' approval. Arguing that the administration was about to lose control of the issue, Train boldly threatened to promulgate unilaterally the regulations. When the White House still did not respond, Train followed through, ending any chance of legislation but ensuring adequate controls. It was another victory for the environment, although, again, Ford's position left little doubt about the sway conservatives enjoyed over administration energy policy.[46]

"[Ford] is now surrounded by doctrinaire anti-environmentalists," Train reflected in his diary. Consensus was difficult on every energy issue, a far cry from the optimism he had held at the outset of the administration. Joining the Clean Air Act as an object of conservative scorn was potential strip mining legislation, another lingering debate from the Nixon years. Train still agreed with moderates in Congress who continued to push for new regulatory authority. Officially the administration had no position on the bill but, behind the scenes, Train remained virtually alone in his advocacy. When Ford called a meeting on the issue in the Cabinet Room, Train entered to find "the regular economic plus energy group," led by Cheney. Initially the group claimed that the strip mining bill in Congress meant an "estimated production loss of 40–160 million tons and a job loss of 40,000." The conversation turned to the rights of the federal government, Cheney and the others arguing that "the States had more and more undertaken effective regulation without Federal legislation." The "guts of the whole issue," Train later claimed, was "East versus West." It was, without saying it, environmentalism versus an extractive economy. Train vigorously questioned the economic costs, continuing to insist that the environment and the economy were not mutually exclusive. If Congress was to pass the bill, he argued, the White House should not veto it. Train did not wilt but neither did his impassioned pleas carry the day. When the bill emerged from Congress, Ford issued his veto. When a revised bill then passed, another veto followed. "With our economy faltering, jobs were crucial," Ford stated. "Since coal was the most abundant energy source this country had, I thought the environmental price tag was too high."[47]

Nuclear power, of course, remained an energy option. When Ford's advisors pushed for increased use with the new "breeder" reactor, Train stalled by recommending consultation with other international powers. "To maximize my credibility," Train later wrote, "I prefaced my remarks by saying I was not an opponent of nuclear power as such." In fact, Train had previously proposed "the phasing out and eventual elimination of all nuclear power as a source of energy on the earth." Waste management issues and problems with plutonium safeguards bothered him, as did potential proliferation of nuclear weapon capabilities. Train was disingenuous but it little mattered. The administration proceeded with its nuclear program.[48]

When Ford's advisors recommended increased leasing of coal on the outercontinental shelf, Train proposed using the proceeds to purchase additional parklands and wildlife refuges. Ford listened politely but the proposal went nowhere. As the end of the year approached, oil policy garnered its own headlines as two large tankers wrecked, creating massive oil spills. As EPA administrator, Train had no direct responsibility but, as he later admitted, this did not keep him from "grandstanding" in advocating reform. Joining with Massachusetts governor Michael Dukakis, Train suggested a system of satellite technology to trace offshore sea traffic. Later he advocated a presidential task force on oil spills and new restrictions requiring double-hulled vessels. Both were expensive propositions and, given industry's influence, both met the same fate as Train's proposals on nuclear power. Train was persistent, if, more often than not, unsuccessful. When it came to energy policy, he simply did not have the clout to back the conservative "doctrinaire anti-environmentalists." Train still saw himself as a loyal Republican and an environmentalist, which unfortunately was an oxymoron for many of his administration colleagues.[49]

With energy policy frequently in the headlines and emboldening his opposition, Train might have avoided newspapers or the television. In fact, events occasionally worked to his advantage, reminding increasingly skeptical Americans that an environmental threat remained. Such was the case when a shocked public learned that the Life Sciences Products Company, under contract with the Allied Chemical Corporation, had polluted Virginia's James River with the pesticide kepone. Several dozen employees were in the hospital as the media reported that the pollutant might spread to the Chesapeake Bay, endangering an array of aquatic life and the surrounding human population. With the Virginia gubernatorial election only months away, Democratic candidate Admiral Elmo Zumwalt charged the Republican incumbent with negligence. All eyes turned toward Train, who was suddenly in demand on television and the radio. When Train acknowledged that EPA had a "problem" and should have stopped the illegal dumping, both the Virginia and Maryland governors tried to deflect criticism onto him. Train answered that the two had taken "out of context quotes" and accused the media of hyping the threat. In a sense, Train had reason for anger. Local reports, at least, sensationalized the event among the nearby community, with Train soon receiving angry phone calls at his Eastern Shore farm. In another respect, however, EPA did share some of the blame and Train's interviews did add to the alarm. It was controversy that cost Train personally even as the outrage he endured reminded the public that industry was not always on the side of the angels nor government regulation always evil.[50]

The kepone controversy paralleled reports that a General Electric Plant had discharged PCBs (polychlorinated biphenyls), a carcinogen, into the Hudson River. When officials closed the river to fishing, media reports compounded a sense of crisis. Once again Train made the rounds of interviews, acknowledging that "the agency should have moved against PCBs sooner." The National Cancer Institute had already begun studies of the carcinogenic effect of over 500 chemicals and, while EPA did not have sole authority to issue a ban, vigilance might have mitigated the damage. It was, according to Train, "one hell of a problem." The public seemed to agree and while the situation again cast Train in an unfavorable light, his loss energized the environmental community.[51]

It all added up to an opportunity, Train recognized. Five years before, in 1971, Train had advocated the Toxic Substances Control Act (TOSCA), legislation to regulate toxic chemicals prior to marketing. The bill had never emerged from Congress, however, facing stiff resistance from the chemical industry, primarily the Dow Chemical Corporation. Using the outrage from kepone and PCBs, Train launched a new campaign to get Congress to act. In a well-publicized speech at the National Press Club, Train employed a bit of hyperbole. It was emotional rhetoric unlike the kind the reserved EPA administrator usually employed but it had its intended effect. Requests for interviews flooded his office, including such shows as National Public Radio's "All Things Considered." Soon the Senate Committee on Commerce held hearings with supporters citing Train's speeches as evidence of the need for reform. With public support, Congress finally reported the bill and Ford, as Train correctly predicted, signed it. TOSCA, a product of Train's persistence, was now law.[52]

Years later Train cited TOSCA as one of his greatest accomplishments at EPA. The legislation was a critical step forward, if industry resistance was any indication of its importance. At the same time, however, the final bill had flaws that Train's praise ignored. The application process was confidential and thus allowed for no public scrutiny and environmental litigation. Moreover, the new law phased out only a small percentage of the PCBs already in use. The National Academy of Sciences reported the law's positive impact but acknowledged that EPA overly emphasized acute environmental effects at the expense of long-term consequences. In this light, TOSCA represented another Train compromise, his praise as much an indication of industry resistance as the actual protection the law afforded.[53]

Democratic Massachusetts Senator Ted Kennedy tried blaming Train and EPA for the persistence of toxic chemicals and pesticides when, in reality, blame more correctly fell with Kennedy's colleagues on Capitol Hill. As the TOSCA debate raged, Train unilaterally banned the chemicals chlordane and heptachlor, an action that angered both the chemical corporations and the agricultural community. "These compounds cause cancer," he correctly declared. "They pose an unreasonable risk to the American people."[54] When the ranching lobby pushed Ford to rescind Nixon's 1972 ban on the use of chemical pesticides to control predators on public lands, Train was vocal—and successful—in his opposition. In a memo to the White House, he defended both the ban and the importance of strong federal management of the public domain. Public lands were for the public interest, he argued, despite "outside interest groups."[55]

The debate over dangerous chemicals on public lands did not garner the publicity of kepone and PCBs but, once again, reiterated that an environmental threat remained. In short, it constituted another opportunity for Train and the environmental community to resume the offensive. With the 1976 election approaching, publicity was sufficient to move Congress to pass two significant environmental reforms that were equally if not more significant than TOSCA. The Federal Lands Policy and Management Act of 1976 (FLPMA) reflected the predator control debate, mandating that the Bureau of Land Management, which controlled the vast majority of the public domain, consider environmental issues among its operational procedures. It constituted a multiple-use mandate similar to that of the Forest Service's and, more significantly, rebutted the burgeoning "Sagebrush Rebellion," a move by western conservatives to turn federal lands over to the states. The Resource Conservation and Recovery Act of 1976 (RCRA) provided an incentive for recycling and recovery of waste materials but also, as a direct reaction to the kepone and PCB controversies, specifically listed hazardous toxic wastes for regulation. These two laws had flaws just as TOSCA did but they also constituted a dramatic improvement in environmental protection. Train surprisingly played little role in either bill, a fact reflected in his memoirs.

He mentions RCRA only once in passing and completely omits FLPMA. One might argue that the latter was not the purview of an EPA administrator. Train, however, had already shown a propensity for stepping outside his assigned duties if he thought it advantageous. Given Train's activism across the board, his absence on these two critical laws—a highlight of the Ford administration—is remarkable.

Train's record in Ford-era legislation was mixed, no doubt a product of both his official regulatory duties and the conservative tide. When it came to water quality, he took every opportunity to stress the need for reform. Invited to a cabinet meeting on the presidential yacht *Sequoia*, Train noted the Potomac cruise "a particularly appropriate occasion" to raise the matter, "especially if we don't have to drink the water."[56] When Congress passed new safe drinking water legislation, Train helped convince a skeptical Ford not to veto. While reports persisted that chemicals found in the water of several large cities might cause cancer, Train shifted gears, assuring an equally skeptical public that regulation was sufficient. "I certainly would have no hesitancy in drinking the water," Train confidently declared—before guzzling a glass as the cameras clicked. Land use legislation, the most egregious failure of the Nixon years, offered no such media moment but did not keep Train from renewing the battle. Predictably, however, the time had passed for such restrictions, Train's pleas carrying even less weight with the growing number of conservatives.[57]

As always, Train found respite from such Washington struggles in his international travels, his forays abroad almost as numerous as during the Nixon Presidency. Just like his predecessor, Ford welcomed Train's endeavors as a form of detente, encouraging his efforts to implement the 1972 Soviet environmental pact and to coordinate NATO's CCMS. Flying literally around the globe, Train visited such countries as Japan, the Soviet Union, China, and Iran, the latter to sign another bilateral environmental agreement. In a trip to London, he enjoyed a long visit with the famous billionaire J. Paul Getty, one of the richest men in the world but then elderly and suffering from Parkinson's disease. In Brussels, Train rode with NATO Secretary General Joseph Luns in an experimental electric automobile, both recognizing a photo opportunity when they saw it.[58] In Moscow, Train was now a familiar acquaintance of Soviet President Podgorny but never forgot that every comment carried potential ramifications. When Train explained that an agreement on migratory birds was proceeding well, Podgorny smiled and exclaimed, "We bring these birds up in the Soviet Union, feed them and make them strong, and then you shoot them in the U.S."

The Soviet leader, it seemed, had an excellent memory; two years before the two had discussed the problem of Siberian snow geese shot in California. Podgorny smoked incessantly and when Train declined an invitation to join him, Podgorny acknowledged that many of his colleagues had given up the habit. "Perhaps I am too conservative," he remarked. "Anyway, I haven't noticed that it has done me any harm." The conversation ended seriously as Train explained that "there is a great deal of real concern and uncertainty among the American people over detente." In many ways, environmental cooperation depended upon "Soviet expansionism in such areas as Africa."[59]

Train remained the consummate diplomat, whether confronting Soviets in the Kremlin or conservatives in the Oval Office. Success depended upon tact, honesty, and the willingness to compromise. The latter was never easy and, when the environment suffered, always frustrating. The result was a mixed record of accomplishment but, Train assumed, it was the way Washington worked. Given the realities of the day, he believed, it was impossible to do better.

In the end, it all came back to the shifting sands of politics and the renewed power of the conservatives. Train did not relish the growing ideological divisions that characterized the nation's political culture and continued to take every opportunity to bridge partisan disputes. He still enjoyed the Washington social scene, maintaining friendships with colleagues from both sides of the aisle. Dinner at the home of conservative columnist Rowland Evans—"Rowly" to Train—still found liberals Ted Kennedy and Walter Mondale, both Democratic senators future presidential candidates, to be regular guests. Train still sent letters of support regardless of party affiliation. He continued to work closely with Senator Muskie, Train's allegiance to the Nixon administration long forgiven. To Democratic Senator Robert Byrd, Train still deserved praise, a "dedicated public servant." Although Lyndon Johnson was the exemplar of liberalism, Train remained friends with his widow Lady Bird, often visiting her at the LBJ Ranch in Texas.[60]

At the same time, however, new Republican minority whip Robert Michel thanked Train for his support, promising that despite Democratic dominance in Congress "we can be heard if we work together." Train's letters to former Nixon administration staffers remained warm. Caspar Weinberger—"Cap" to Train—sent thanks "now that the darkness has pretty much disappeared." Although he had long clashed with Maurice Stans over environmental policy, Train sent his condolences when Stans pleaded guilty to violating campaign finance laws. "Your letter was a lovely expression of warm friendship," Stans replied.

To Train, policy differences did not mean personal animosity. Reasonable men could still work together to cross the divide, for the good of the country if not their cause.[61]

Train continued as he always had, enjoying his many friendships. At a dinner party that included noted pundit John McLaughlin, Supreme Court Justice Lewis Powell, and a number of foreign diplomats, Train boldly asked actress Elizabeth Taylor to dance. "I felt like a sweaty high school adolescent," he recalled, flustered enough to mumble to the star, "I can't wait to tell my twenty-year-old son about this." When he asked where she came from, Taylor replied, "The moon!" Train, a man accustomed to the rich and powerful, had, it seemed, met his match. Dignitaries continued to arrive at his Eastern Shore farm, which now sported a large red Southern Railroad caboose in the yard. "I guess that if your name was Train you'd want one too, wouldn't you?" answered Train to an inquisitive journalist. "You get the feeling that Russell Train has been kidded about his caboose by experts," the reporter wrote. "You also get the feeling that this is a man who cares about what happens to America." It was one sentiment, at least, on which some Democrats and Republicans could still agree.[62]

The impending presidential election, of course, did little to help Train in his bipartisanship. Both conservatives and liberals pressed their cause in the hopes of electoral gains. "There is an increasing sensitivity in the Congress to public reaction against big government and regulation," Train wrote after one cabinet meeting. In another meeting Train sat next to fellow moderate George Bush, then CIA director. "Did you notice that the first three bills on the Administration's 'don't want' list were all EPA's?" Train whispered. "Yes," replied Bush, "you are about as popular as the CIA." White House staffers continued to joke with Train but their humor also carried a hint of sarcasm. "I don't know if I should sit next to one of the bad guys," laughed one aide as he pulled up a chair next to Train. EPA, another cracked, "needed emission standards for the candle lobby." Train answered with his own wit. Faced with a sarcastic question about whether couples should file environmental impact statements before having children, Train responded, "I doubt very much that an EIS could be completed in the requisite nine month period."[63]

The best retort, of course, was to remind party faithful that he was no liberal. Train indeed applauded moderate restraints on taxes and spending as a correction to the Great Society's excesses. While remaining cordial with Supreme Court Chief Justice Warren Burger, Train privately thought the famous broad constructionist "a very silly, vain man." In one instance, Train silenced potential critics with his own proposal for a flat tax. Published in a long *Washington Post* article, Train's "plan for real tax reform" cited the complexity of the tax code and claimed "the present tax system helps erode the credibility of the government." The proposal won praise from conservative columnist James K. Kilpatrick even as many Democrats argued that potential revenue would prove insufficient. In a sense, the proposal was astounding given Train's record of promoting environment-related tax incentives and it constituted a complete break from his own past. Years later Train denied that the flat tax article was a conscious attempt to curry favor with the growing number of conservatives. Neither, however, did he acknowledge that the proposal was a direct contradiction of his own record.[64]

In reality, neither Train's flat tax proposal nor his professions of party loyalty kept him off the defensive. "Ecologists are the true conservatives," he insisted at every opportunity as the election approached. Routine hate mail arrived at his office. "You have a brain like a jackass," wrote one critic in a screech against government regulation. "Insomuch as EPA has come to be staffed entirely with assholes," another anonymous letter read, "it should be abolished by Congress."[65] The most troubling letter came from a California prison. Lynette "Squeaky" Fromme, one of the Manson girls serving time for an attempted assassination of Ford, threatened Train's life. "You weak mealymouth," Fromme wrote, prompting Train to request an armed guard. Train somewhat overreacted, but the entire episode seemed to represent perfectly a man—and his cause—under attack.[66]

The real threat came from the presidential campaign of Republican Ronald Reagan. The charismatic California governor, a former actor and Democrat, rallied conservatives in his attempt to wrest the Republican nomination from Ford. Stealing the nomination from an incumbent President was a tall order but, in the early primaries of 1976, looked like a strong possibility. As late as June, only weeks before the Republican convention, Rowland Evans lunched with Train and predicted the nomination too close to call. Ford, he claimed, "had never taken Reagan seriously enough." The nomination was "fifty-fifty." Train anticipated a close vote, each Ford primary victory a "relief." At the same time, however, Train saw little evidence of White House overconfidence. Ford's drift rightward accelerated in a clear attempt to pacify the conservative insurgency.[67]

The campaign weighed heavily. Rockefeller, Train reported, was "down." Sounding defeated, Rockefeller made comments that seemed almost a plea. "I am doing the best I can Russ," he mumbled. "I just don't know." Goldwater had largely kept his promise not to attack Train personally but had become "less than cordial." He hated EPA, Train assumed, "now considering me practically a communist."[68] At every turn conservatives appeared energized, a phone call from the White House only one troublesome example. Senator John Tower, Rumsfeld explained, had requested that Train fill one of the deputy EPA regional administrator positions with a lower-level employee. The individual was less qualified than Train's own choice but had worked personally with Tower and had befriended the powerful Texas conservative. Tower had promised to serve as Ford's floor manager at the Republican National Convention and threatened trouble if Train failed to make the appointment. Annoyed over the flagrant imposition of politics into the management of a regulatory agency, Train leveled his own threat. He would make the appointment but then resign, an option that promised the White House additional negative publicity in the heated campaign season. After an angry confrontation with Tower in the Senator's "hideaway" office in the Capitol, Train offered a compromise—a third candidate that no one had proposed. This candidate had sufficient credentials and thus Train felt vindicated, although the entire episode was "unpleasant" and a reminder that the traditional ways of Washington were rapidly fading.[69]

When Rockefeller withdrew as Ford's vice presidential nominee, the selection of his replacement assured intense competition. Rumsfeld, now Secretary of Defense, had hoped to win the post. He was, Train noted, "an ambitious guy." Train's friend Elliot Richardson was a strong possibility because his comments and actions had, like Ford, drifted to the political right. The "new conservative Richardson," as Train characterized him, now frequently gave "free enterprise, anti-government regulation speeches." Train recommended several candidates to the White House, a list led by moderates—and friends—William Ruckelshaus, George Bush, and Howard Baker. Predictably, the recommendations did not sway Ford, whose selection of Kansas Republican Robert Dole was, according to Train, a "disappointment to those of us who are liberal or moderate Republicans." In the end conservatives promised a floor fight were the convention to nominate a moderate. Reports indicated that Ford had planned to select Ruckelshaus but that conservatives had insisted upon a southerner or westerner. Ford, however, denied that appeasing conservatives was his sole criterion and claimed that Dole's appeal to the "farm states" played a role as well.[70]

If Dole's selection did not indicate how seriously Ford took the Reagan challenge, the party's platform and convention demonstrated the GOP's ideological drift. Train had long pushed for a strong environmental plank, writing Ford on one occasion that "the Republican Party had a great conservation record but was rapidly throwing it away." When it came time to draft the platform, however, no one contacted Train "for any substantive input whatsoever." The result was a platform with only passing mention of conservation. It promised multiple use "where compatible" and public lands to be open to mineral development, at least in the absence of overriding interests to the contrary. "It certainly does not represent a strong or even a positive environmental plank," Train acknowledged.[71] As the convention approached, Train received a letter from Cheney promising him a role in publicity. When details did not follow, Train called the campaign to inquire. Cheney's letter, came the reply, was a mistake. "You are not expected." It was somewhat of a shock; Train had, after all, taken part in earlier conventions. Upon reflection, however, the reason was obvious. "My identification with the administration is not particularly considered an asset," Train wrote. "The way my noninvitation was handled hardly added to my enthusiasm for my party."[72]

The convention went smoothly. Relations between Ford and Reagan remained strained but the party put forth a united front despite a significant split in delegates. Watching the festivities and Ford's ultimate nomination on television at home, Train reflected. "I felt oddly out of the picture," he wrote. "I am not really in tune ideologically with the position the party is taking today." Predictably, Train played little role in the general campaign against the Democratic challenger, former Georgia governor Jimmy Carter. When the Democrats questioned the administration's implementation of environmental policy, Train became defensive and called a press conference to answer the charges. Despite it all, he still saw himself as a "team player." For those who cared about the natural world, however, Train's silence on the administration's overall environmental record spoke volumes.[73]

Carter, by contrast, had a fairly strong environmental record. He ran a solid campaign that reflected environmentalists' concerns and appeared ready to solidify the Democrats' grip on the constituency. While governor of Georgia from 1970 to 1974, Carter consolidated most of the state's conservation functions into a Department of Natural Resources, put a vigorous administrator in charge, and secured budget increases. As a presidential candidate, he promised to sign the strip mining legislation that Train had advocated and that Ford had twice vetoed. He promised to slow down the construction of dams and other reclamation projects and further protect wildlife and wilderness areas. Most importantly, his language surpassed Ford's: "I tell you that this is no time for those of us who love God's earth and the beauty of it, the purity of air and water, to compromise or to retreat or to yield in any possible measure to the devastation or deterioration of the quality of our lives or our environment."[74]

Train had met Carter the year before at the Southern Governor's Conference and, indeed, Carter's environmental interest seemed genuine. Reflecting the growing conservatism of the South, the majority of the governors present complained about EPA and federal mandates, Train being the proverbial punching bag. Carter did not take part in the grilling but cornered Train afterwards for a long talk on the importance of environmental policy. When Train suggested that Carter join the other governors, Carter replied that their conversation was more important—a sure way to win Train's admiration. Now, during the presidential campaign, Train clearly wanted to remain a team player. He had along allegiance to the Republican Party and remained proud of his record. At the same time, however, a number of EPA's professional staff quietly planned to vote Democratic. Former Oregon governor and fellow moderate Republican Tom McCall planned to do the same, much to Train's surprise. Train found himself deflecting questions about whether he would continue to serve were Carter to ask, many of his colleagues aware of both his frustrations and Carter's record. In short, a potential Carter Presidency did not promise an environmental Armageddon even if it did threaten Train's own government service.[75]

The Carter campaign was adept. It accelerated the migration of environmentalists to the Democratic Party without completely alienating conservatives. Carter tailored his comments to individual groups and remained vague when necessary. Doing so, Carter successfully bypassed the inevitable questions of energy versus environment. "I don't see myself as a liberal or conservative or the like," he claimed. "If [voters] are liberal, I think I'm compatible with their view. If they are moderate, the same; and if the voter is conservative, I think they still feel that I'm a good president." Carter was a southerner, after all, a man who claimed a deep faith in traditional values and, as such, attracted social conservatives. He was a farmer, a veteran, and a businessman—a regular guy.[76]

The election of Jimmy Carter as the nation's thirty-ninth chief executive on November 2, 1976, was not a repudiation of conservatism. For one, the election was close. In the electoral college, the vote was 297 to 240, the smallest winning total since Woodrow Wilson's victory in 1916. Only 1.7 million popular votes separated the two candidates. It was, however, more than this. Ford had trapped himself. His pardon of Richard Nixon soon after assuming the Presidency irrevocably tied him to his disgraced predecessor, a political albatross. Carter knew as much. "The most significant factors [for the victory]," Carter later wrote, "were the disillusionment of the American people following the national defeat in Vietnam, the Watergate scandal, and my success in convincing supporters that we should keep our faith in America and that I would never permit any repetition of such embarrassments."[77] In other words, the victory owed less to larger questions of government and regulation than matters of trust and character. In fact, the conservative movement continued to grow even as Ford garnered little enthusiasm among its ranks, his many efforts to the contrary aside. With each passing month, conservatism appeared more popular and more Republican. Train well knew the growing power of conservatives, which Carter soon discovered as President.

In many respects, as 1976 wound to a close, Train felt a rush of emotions. He loved his job but knew that his support of Ford precluded any significant role in the new administration, despite speculation to the contrary. A certain melancholy descended upon administration personnel, who were busy in the transition and packing their belongings. Train's letter of resignation reflected this melancholy, exhibiting both a pride in his accomplishments and a realization that the task was far from complete. "It is my expectation," he wrote, "that I will continue to work for this goal as a private citizen." Train assured a smooth transition, fielding calls from exuberant Democrats and encouraging fellow Republicans to cooperate fully without simply "marking time." Later, new EPA employees congratulated Train and thanked him for leaving "no booby traps." Letters of support continued to arrive along with praise from the media. Train had long loved the attention, with such laurels undoubtedly blunting some of the sting of his final departure.[78]

Even as Train asked Ford to sign a photograph of the two together, anger was not far below the surface. In what must have been déjà vu to Train, Ford denied his pleas for an increase in EPA's budget. The agency desperately needed the funds but Ford, like Nixon, ended his Presidency with a conservative outcry. Privately, budget officials described the action as a political vendetta, forcing the incoming Carter administration to recommend budget increases. During the campaign Train had held his tongue but now, with less to lose, he lashed out as well. "The hardest part of the job still lies ahead," he declared, "with EPA's needs critical." Despite his many successes, he felt "jaded and battered."[79]

His anger was not just over a sense of betrayal. "If a decision doesn't go as far as our environmental friends would like, it is immediately called a sellout," he stated in another interview. "If a decision goes against industry, we're accused of giving in to environmental emotionalism." Train continued to insist that he was the true conservative. He did not embrace "the traditional liberal reformer's approach which is 'top down'" and "never had any intention to force [regulation] down everyone's throat." Washington just appeared so open to extremes. Now, it seemed, no one was willing to compromise.[80]

After eight years in government service, Train had learned to walk a "fine line."[81] Now, facing an uncertain future, he was both hopeful for Carter's Presidency and saddened for his own career. He was proud of his accomplishments and angry at the shifting political tides in the nation's capital. He remained a loyal Republican but appeared relatively powerless in its evolution. Returning to the private sector, Russell Train was clearly a man conflicted.

9

Wildlife and Washington

It took only weeks for Train to find the silver lining in the black clouds of Washington. Retreating to his farm with no official duties, Train sat with Aileen before a large fire. There it dawned on him; he had no paper or reports to read, no speech to prepare. "It is a marvelous feeling," he reflected in his diary. For weeks his allergies had worsened and he had endured painful lower back spasms, physical symptoms that appeared to mirror his mood. Now, however, "I must say that it is good not to be under that kind of pressure any more."[1]

The future was not necessarily bleak but rife with possibilities. The Conservation Foundation offered a senior fellowship, a largely symbolic post that nevertheless provided an office of three rooms and a "place to hang my hat." Colleagues and acquaintances arrived with more permanent and substantive offers. He might run the Alliance for Saving Energy, for example, or the U.S. Association for the Club of Rome. He might lead the citizens' advocacy group Common Cause, an organization George Bush described as "predictably pro-Democrat and left-wing" and recommended against. With the Train's youngest child, Errol, preparing for college, a new home seemed to symbolize the change in Train's life. Selling their home on Woodland Drive, the Trains purchased a large, four-story townhouse condominium in the prestigious Kalarama area of northwest Washington. Purchased with over $200,000 in cash and only a three-year note for the remaining $160,000, the new home represented a new beginning, Aileen quickly hosting a cocktail party for their new neighbors.[2]

Most encouraging, Train now recognized, was a new freedom. No longer serving a policy master, he was able to embark unobstructed on an agenda of his own. The result, predictably, demonstrated his long-standing faith in moderation and compromise as he worked to unite disparate interests on behalf of the environment. He agreed to serve on the boards of both the Natural Resources Defense Council (NRDC), one of the most important public interest law organizations, and the Union Carbide Corporation, perhaps the most prominent chemical company in the world. The latter had "many environmental problems," Train acknowledged, including an array of pesticides and other toxic substances. It seemed an odd choice but, to Train, represented an opportunity for improvement. Convinced that his appointment was not simply window-dressing and assured that he would not have to lobby EPA, Train willingly accepted the $25,000 bonus the position carried. Train later denied that serving two organizations on opposite extremes of the environmental spectrum was a conscious attempt to forge a middle way, yet his acceptance arguably reflected his overall philosophy. Years later, after a tragic and deadly accident at Union Carbide's plant in Bhopal, India, Train led efforts to assure corporate safety and environmental protection.[3]

In the interim, the lecture circuit allowed him to hit two birds with one stone. Commanding speaking fees as high as $5,000, Train found his old gospel of environmental protection lucrative in a way unimaginable before. Traveling the country continuing to spread the environmental alarm, Train agreed to serve on the National Academy of Sciences' World Commission on Coal Policy. Once again the choice reflected his philosophy. While many environmentalists argued against coal, Train insisted that his service on the commission might encourage better coal technology. He knew that coal was the world's most abundant and cheapest energy source and that its continued use was inevitable. Like many of his colleagues, he too readily dismissed coal's implications for global environmental change—the "one imponderable," he mused. Nevertheless, his efforts represented a realistic alternative to the nation's needs. Two years later, presenting the commission's report to Carter in the Oval Office, chairman Carroll Wilson of the Massachusetts Institute of Technology insisted that the recommendations, if adopted, represented adequate environmental protection. "Do you agree?" Carter suddenly asked Train, who answered affirmatively. "That's all I need to know," Carter replied, which was both a compliment and a recognition of Train's stature.[4]

Train's reputation brought additional opportunities to bridge environmental divisions. At the behest of the Sierra Club's Michael McCloskey and the Atlantic Richfield Corporation's Robert Anderson, Train agreed to head the new organization RESOLVE. Dedicated to mediation rather than litigation, RESOLVE sought nonadversarial solutions in the environmental arena, which to Train was "an obvious reflection of my personality and career." Traveling to meetings in San Francisco and Texas, Train and his colleagues counted several minor successes but hardly stemmed the growing tide of environmental lawsuits. Within two years, Train negotiated a merger with the Conservation Foundation, which took over RESOLVE's functions and largely ended its independent role.[5]

A more obvious success was Train's role in mediating a long-standing dispute between the environmental community, EPA, the New York State Power Commission, and several utilities led by Consolidated Edison of New York. Given growing energy demands, ConEd planned to build a large "pumped storage" facility at Storm King Mountain along a scenic stretch of the Hudson River. There it planned to accumulate water during periods of low energy demand, releasing it through power-generating turbines when necessary. In addition, ConEd and the other companies regularly pumped water to their operations, returning the water to the river once their plants were cool. The resulting warm water killed thousands of fish and angered the Hudson River Fisherman's Association. To EPA the solution was the installation of cooling towers in lieu of the river water, a proposal that brought howls from nearby communities, which understandably recognized the potential blight. Regularly flying to New York, Train learned how "acrimonious" discussions had become. With EPA threatening to pull out of mediation and the utility companies "smirking"—according to environmentalists—Train "managed to hold it together" by asking the utilities to investigate changes in plant operations and the use of fish hatcheries. Meetings followed and months passed before Train reported "an important breakthrough in my negotiations." ConEd agreed to surrender its Storm King plans and donate land for a large public park. In exchange, it would not have to build the expensive cooling towers provided that technological and operational modifications mitigated the fish kill. A new organization, the Hudson River Foundation, promised to monitor the health of the river and provide a yardstick for compliance.[6]

With ConEd providing most of the concessions, the environmental community applauded and congratulations flooded Train. The *New York Times* ran a picture of a smiling Train bending over a somber Charles Luce, ConEd's chairman, as Luce signed the agreement. "All participants agreed," the paper reported, "that no settlement would have been possible without the mediation of Russell Train." Reports cited Train's "rare prestige and skills," while one colleague wrote that the agreement was "worthy of a Nobel Prize." Train acknowledged the settlement was "one of the most satisfying events in my life" but also sought to use the occasion to encourage mediation and compromise in other areas. Sending copies of the agreement to his environmental colleagues, he declared it an excellent example. "This demonstrates dramatically to the entire nation that environmental and energy needs can be effectively balanced."[7]

Operating essentially as a free-agent environmentalist, Train assured that his speeches and mediation did not prevent travel, one of his first loves. With more time for hunting, he visited Long Point, where he shot ducks with the manager of Harvard's billion-dollar endowment, among other notables. He cruised the Caribbean with friends aboard the *Condie,* a forty-eight-foot yawl that Train described as "close to perfection." Swimming before breakfast, fishing for barracuda, and watching beautiful sunsets with scotch and vodka in hand made the frustrations of Washington a distant memory. Environmental advocacy brought Train to England, Germany, and Saudi Arabia, the latter particularly memorable for Train. Invited by Earth Day organizer Denis Hayes as part of a group of private environmentalists, Train worried that the Saudi government might use the group for propaganda. With two of the delegation Jewish, Train stressed a "just and lasting peace in the Middle East" as a "prerequisite for environmental protection." Hosted by Prince Saud al Faisal, Train enjoyed "lavish dinners" as he mingled among the royalty. For a man who had seen much of the world, he found Arabia surprising. Rain was frequent, while a visit to Asir Kingdom Park exhibited "terraced valleys, magpies, and brilliant blue lizards sunning on rocks." It was "spectacular but wholly unexpected." Afterwards, Train and his colleagues advised their hosts to adopt a "strong, explicit statement of national environmental policy." The group returned in time for Train to watch the famous David Frost interviews of Nixon, the former President's first public overture since his resignation. "It was pretty ghastly," Train concluded. "Nixon doesn't become more sympathetic or appealing with time."[8]

The days passed quickly, the back spasms lifted—and the restlessness returned. Refreshed, Train could not simply enjoy the good life for long; his innate drive and ambition would not let him. He felt he "lacked focus" and, somehow, wanted more. The sudden death of his brother Cuth from a heart attack reinforced the "transitory nature of your own life." Anxious to return to the environmental struggle full-time, Train found an offer from his friend, S. Dillon Ripley, to lead the World Wildlife Fund–United States (WWF) "more than a little interesting." Ripley, secretary of the Smithsonian Institution and chairman of WWF's board, insisted that his colleagues "in the vineyards of international conservation" were unanimous in their opinion of Train. He had "an impeccable reputation" and all the necessary connections. Present employee Tom Lovejoy could handle the scientific aspects of the job, leaving Train to do what he did best—public relations, lobbying, and fund-raising. Train could set a new agenda for the organization, take it to the next level. "I urge you to do this," Ripley solemnly concluded, "for the interests of conservation."[9]

The offer looked like a perfect fit. Just as in the past, Train's extensive connections proved pivotal; a lifetime of cultivating friendships in the political and environmental communities paid dividends. Train had already proved himself a formidable fund-raiser and his track record in international wildlife was obvious. To many environmentalists, WWF had too high an administrative cost per dollar raised and was both too elitist and too social, "too many princes and polo players" in the words of Laurance Rockefeller. This was Train's world, however, as it was Rockefeller's. Train operated easily among the power elite and, in the least, hoped to improve the effectiveness and reputation of the organization. The group's headquarters was in Washington, Train's base, and, like Train, it traditionally shied away from confrontation. Its old-boy, connected background assured few radicals. Concentrating abroad, Train could choose his domestic battles wisely. Aileen, who already served on the board, urged her husband to accept and Train agreed. Demanding a salary just as he had with the Conservation Foundation thirteen years before, Train became the WWF's first paid president and chief executive officer. It was, he wrote in his diary, "a new milestone."[10]

Along with the new job came news that Train had won the prestigious Tyler Prize for Environmental Achievement, established by John and Alice Tyler of Los Angeles and carrying a hefty reward of $150,000. Flying to Pepperdine University to accept the award before a large crowd, Train was "pretty emotional." It was a "very conservative crowd," Train reminisced, "and I am not at all sure that Pepperdine doesn't feel some embarrassment in front of all their oil and other business friends seeming to be supporters of ecology." Pat Boone sang the national anthem, Train amazed that "he looks about twenty-two and is a grandfather." In his acceptance speech, Train returned to a familiar theme—the importance of "non-adversarial approaches to the resolution of environmental conflict."[11]

Train appeared to be on a roll but was still open to surprise. Accustomed to the media glare, he had grown to love the limelight. To him the new job and award was just another step in his illustrious career, and surely of interest to the public. The press, however, seemed not to care. No media attended the award ceremony, causing Train to conclude, "The PR aspects of this trip have been a disaster." Worse, the press conference to announce his new job drew only one reporter, Train's former CEQ press officer then working for the *San Francisco Chronicle*. Train attributed the lack of interest to the previous night's snow but remained somewhat annoyed. "The award seems to be the best kept secret of the year," he wrote.[12] It was a lesson he could not escape; his new job was important but, out of government, he had lost access to his bully pulpit.

Train wasted little time. WWF drew less than $2 million per year from only 30,000 members. With just twelve employees and the resources of the richest nation in the world, it drew frequent criticism from other nations' WWF affiliates. Primarily a grant-making agency, it ran few projects of its own, most focused narrowly on the conservation of specific species. To Train, the budget was "inflated in projected income." Because the staff was too timid to level with the board, the organization committed to grants it could not afford. Train quickly closed the San Francisco office because it "had accomplished nothing" but raised staff pay as a morale boost and as an incentive for recruitment.[13] He launched a new direct mail campaign, coordinated with the Book of the Month Club, and convinced the famous author James Michener to promote conservation in the Chesapeake Bay area. Identifying the ten most important donor states, he ordered direct appeals to their residents. Publicizing the organization frequently required humor. After listening to Train describe the threat to wildlife, one potential donor asked what he could do. "Give us money," Train half joked, "as much as you can."[14] Fund-raising lunches and dinners soon dotted Train's schedule, at least temporarily causing him to give up drinking. In between, he wrestled with meetings, phone calls, and letters.

Train's impact was immediate and obvious. In his first year, WWF's income rose 35 percent, with Train's new corporate campaign more than doubling its revenue. The organization moved into new spacious offices on Connecticut Avenue, occupying the entire eighth floor of a large office building. While continuing to fund a number of organizations, Train cancelled key grants to others, including the Nature Conservancy. WWF expanded its own programs, first in Central and South America and, later, in Asia and Africa. Projects included Brazil's coastal forests, Costa Rica's parks, India's tiger habitats, a number of initiatives in Madagascar, and a global primate program, among others. Reflecting the same evolution he helped engineer at the Conservation Foundation, Train encouraged broader ecological approaches to wildlife conservation. This meant WWF had to consider plants and animals "in the context of other urgent human needs, such as agriculture." Ecosystem protection was necessary but WWF, like Train, was not radical.[15]

Years before, Train had led the official American delegation to the initial meeting of the Convention on International Trade in Endangered Species of

Wild Fauna and Flora (CITES). Now Train led a large contingent of nongovernmental organizations (NGOs), covering his rather aggressive assumption of this role by joking that he "had been elected by a Democratic process that would not bear examination." At this 1979 Costa Rican convention, Train succeeded in winning for WWF control of the newly formed TRAFFIC program. Part of a network centered in England, TRAFFIC monitored international trade in wildlife through American ports and reported problems and illegalities to the government and other environmental organizations. WWF, Train argued, would provide funding and its headquarters in the nation's capital would prove beneficial. With TRAFFIC, WWF became the main NGO for implementing the convention's agreements, distinguishing itself from the other twenty-six environmental organizations present. While at the convention, Train met Costa Rican President Rodrigo Carazo and toured the country's largest national park, admiring over fifty species of birds.[16] Later that same year Train chaired the steering committee for the World Conference on Sea Turtles Convention in Washington. "Let us determine that this is not going to be a conference simply for the sake of a conference," he declared in his opening remarks. Train's efforts assured that it was not. The conference identified 140 specific projects around the globe to protect the endangered turtle's migratory routes and beach nesting areas. Annoyed at the resistance of Japan, Train convinced the delegates to support the strong CITES ban on turtle-related products. For its part, WWF paid local inhabitants to mark, enclose, and protect critical turtle habitats.[17]

For years WWF had supported the International Union for the Conservation of Nature and Natural Resources (IUCN). With its contributions increasing, IUCN elected Train its vice president for the North American region. Predictably, the responsibilities increased Train's travels. Flying to board meetings of IUCN and the international coordinating body of WWF, attending conferences, and monitoring WWF programs brought Train to every corner of the globe. A trip to the Galapagos Islands with former Nixon colleague Elliot Richardson fascinated Train no less than it had Charles Darwin generations before. Following in the Darwinian tradition, Train and his family rented a boat named the *Beagle*. Despite a stomach ailment, Train witnessed a volcano eruption and rare birds such as the black-billed cuckoo. "Tropical birds with their long streaming tails and brilliant red beaks flew over and around the *Beagle*," Train recorded, "making their clattering call as they went." The threat from tourism and a growing population was obvious, and Train assured WWF financial support for an array of environmental initiatives. An interesting sidebar to the trip was a discussion with Richardson about former Attorney General John Mitchell's role in the Watergate scandal. Despite historical accounts that made Mitchell one of the masterminds, Richardson insisted that Mitchell was "essentially naive," too inexperienced to halt the excesses of subordinates. Mitchell said nothing in meetings, only appearing wise at the end by summarizing what others had said. Nixon kept White House tapes, Richardson also surmised, because of his disappointment upon reading Robert Blake's biography of British Prime Minister Benjamin Disraeli. The book missed Disraeli's true motivations and character, Richardson explained, and Nixon wanted to ensure that future biographers had sufficient documentation to grasp his own greatness. Richardson knew Mitchell and Nixon well, of course, and Train thought his comments "a fascinating hypothesis."[18]

Aileen was Train's most important travel companion, joining him no matter how arduous the journey. In a trip to Brazil to study the importance of the jungle's many plants to the health of the Amazon basin's aquatic life, the two ventured by open boat into the proverbial heart of darkness. Camping on huge floating logs tied together between a small island and the mainland, they enjoyed a meal of piranha. With no sign of the modern world but the jungle teeming with life, Train acknowledged it "an eerie sensation." Equally impressive was a trip to Michoacán in Mexico. There they saw millions of monarch butterflies clustered on trees, the culmination of their annual migration. "We lay on our backs in a grassy clearing and gazed at the countless butterflies silhouetted against the sky overhead," Train wrote. "The constant beating of their wings sounded like the movement of wind through tree branches or the fall of raindrops on leaves." Predictably, the Trains' trips resulted in significant WWF aid to both areas; human expansion was threatening both the fragile Amazon ecosystem and the famous monarch migrations.[19]

As the late 1970s passed, Train's responsibilities kept him in constant contact with the famous and powerful. In one trip to England, a cocktail reception at 10 Downing Street led to a long discussion with Prime Minister Margaret Thatcher. She was "very direct and forceful," Train believed, a point on which her political opponents undoubtedly concurred. In a trip to Nepal, Train met Mother Teresa at the British ambassador's Kathmandu residence. Train praised her but challenged her opposition to contraceptives. The famous minister to the poor of Calcutta replied quietly that the "temperature method" was best. Monitoring one's body temperature for ovulation, Train tactfully kept to himself, was impractical for the masses of ignorant and poor. At the German embassy in Washington, Train sat next to reporter Barbara Walters, "Alan Greenspan's girl." At the

Japanese embassy, he talked with Carter's son Chip and Chip's new bride Caron. The latter's public references to Carter—the President of the United States—as "Jimmy" struck Train as rather odd.[20]

The Washington scene brought endless cocktail parties, dinners, and tennis games. Dinner at the home of *Washington Post* publisher Katharine Graham was "pretty heavy going." Perhaps a bit unnerved by such a forceful woman, Train acknowledged, "I usually feel pretty intimidated by Kay." Oddly, political discussions with Robert McNamara, one of the key architects of America's Vietnam policy, and Paul Nitze, famous proponent of containment, were less stressful. Enjoyable were tennis matches with humorist Art Buchwald. "Art makes up for his lack of tennis skills," Train remarked, "by the high quality of his entertaining."[21] One summer vacation sparked a close friendship with famous CBS newsman Walter Cronkite, who owned a beach house nearby. Soon Cronkite invited the Trains for sailing on his forty-six-foot yacht, the guests snacking on stuffed eggs and lobster sandwiches while drinking white wine. Always polite and an interesting conversationalist, Cronkite could almost match Buchwald's humor. With drinks flowing at one party, Cronkite did a "fantastic mock strip tease to the Sugar Blues." Train agreed with columnist Ann Landers that Cronkite's mass appeal owed much to his "middle American 'gee whiz' quality."[22]

Politics, of course, was never far below the surface. The jokes and drinks did not always disguise the serious business around which almost all of Train's Washington circle of friends revolved. In any event, the Carter administration certainly enlivened conversations. Even as Trail experienced tremendous success at WWF and relished the freedom from his personal frustrations at EPA, Carter's environmental policy faced the same impediments that Train well knew. Train and his environmental colleagues had held high hopes but the talk at parties soon turned to "Washington's anti-Carter feeling." Most Republicans encouraged such sentiments, of course, but many Democrats, whose hopes had far surpassed Train's, found Washington's reality particularly bitter. "There is great criticism of Carter from Democrats," Train wrote early in 1979. "I was surprised by the strength of negativism and found myself standing up for him."[23]

Train could certainly empathize. Like Train, Carter accomplished much on behalf of the natural world. In the end, Carter's record is solid. At the same time, however, American conservatism continued to grow, as did the partisanship surrounding the environmental issue. Reeling from the Nixon and Ford retreat and encouraged by Carter's rhetoric, environmentalists demanded much.

Carter faced worsening energy and economic problems, however, which forced him into significant compromises. In 1979, authors Edwin Diamond and Bruce Mazlish claimed that Carter's policies were "blurring the lines between liberalism and conservatism." The following year author Lawrence Shoup claimed that Carter was similar to many Republicans.[24] Such sentiments underestimated Carter's essential liberalism, the creation of another large federal bureaucracy, the Department of Education, as only one example. Nevertheless, in many respects swimming to the political left placed Carter against strong tides in an increasingly divided nation's capital. In the words of historian Jeffrey K. Stine, "Carter inevitably fell well short of meeting the environmental community's unrealistically high expectations."[25]

The economy continued to embolden fiscal conservatives. By 1979, 70 percent of Americans viewed the economy as the nation's most important issue. With the gross national product growing at a paltry 1.8 percent, less than half the average for the previous two decades, inflation and unemployment rose. "The shriller you are," admitted Terry Dolan of the National Conservative Political Action Committee, "the easier it is to raise funds."[26] Just like many environmentalists, social conservatives had held high hopes for the Carter Presidency. Carter's policies on issues such as homosexuality, abortion, and the Equal Rights Amendment, however, further drove them into the GOP camp. Jerry Falwell, founder of the Moral Majority, led this migration even as he described the environment as an "overblown issue" that liberals used to control American lives.[27]

"Militant environmentalists," Train publicly acknowledged, just did not understand the political tides. The key was not confrontation and a stronger alliance with the Democrats but rather a bipartisan approach that stressed compromise. The new administration, Train predicted to the *Washington Post* two months after Carter's inauguration, might succeed on many "small incremental issues." Rancor, however, would preclude any "great new environmental initiatives." In one NRDC board meeting, trustees "discussed the environmental backlash and what to do." Several members recommended an expensive lobbying and public relations campaign aimed against the new conservatism. This would prove counterproductive, Train argued, and encourage partisan legislative gridlock.[28]

On the other side, Train found many of his friends equally exacerbating. "A number of Republican meetings were being held," Train wrote in his diary. "I have a feeling that when all is said and done, I might not want to be associated with the product." Train continued to contribute to Republican causes and found much of the Carter White House distasteful. Carter's top aide Hamilton Jordan, for one, "wears old clothes to the office with no tie." He "plays tennis at Arlington in the early afternoon" and is "apparently sleeping with a black secretary at the White House." Such behavior detracted from the dignity the administration needed. "It is hard to see how Carter is going to pull his act together with that kind of staff support." At the same time, however, Train lamented that his friends "are getting more conservative"; there appeared "little middle ground." As one example, Rowland Evans was "consistently anti-environment and pro-nuclear." In short, "there just does not appear to be any [moderate] candidate." Both Democrats and Republicans had their partisan knives sharpened, with Carter—and hopes for environmental policy—caught in the middle.[29]

True to his creed, Train avoided the shrill attacks from both extremes and kept in contact with the new administration. He developed a particularly close relationship with his successor at EPA, Doug Costle. Upon learning that he was Carter's choice to lead the agency, Costle called Train to arrange a lunch meeting. Train enjoyed playing the role of environmental sage and quickly dispensed advice. The two met periodically at Costle's office, Train's home, and official ceremonies, while Train "put together some interesting and timely material" when he thought a point particularly salient. "He seems on a good, sound track," Train concluded. Other administration officials were less impressive. Charles Warren, the new CEQ chairman, was a "very nice guy but without much cutting edge." His failure to call for advice was especially irksome. "I thought it would be natural for him to contact me," Train privately complained. Warren's 1979 replacement, Gus Speth, was an improvement, at least in the sense that he showed proper deference. Early in Carter's term, Train invited new Secretary of Interior Cecil Andrus to dinner. Lavishing praise, Train described him as the best candidate for the position. An amicable relationship then developed, although privately Train worried that Andrus wanted control of EPA.[30]

This was, after all, a real possibility. Proposals for Executive Branch reorganization included moving EPA into the Department of Interior. Other proposals included abolishing CEQ or moving it to EPA. After Train testified against reorganization before the Senate Public Works Committee, Costle visited Train and asked him to continue lobbying individual senators. Costle was worried about his job, Train wrote: "Doug is still anxious that he not be identified with any effort of this sort." Andrus, for his part, promised in a meeting with Train to oppose reorganization, although adding that if Carter insisted, "I would be a good soldier." This promise of opposition, Train thought, was not "entirely ingenuous." If a cabinet secretary publicly opposed reorganization of his own department, it "would be inconceivable" that a President would proceed. "I think Cecil is dying to become head of EPA," he wrote. Shaking hands with Andrus as their meeting ended, Train promised to fight publicly the proposal—"but nothing personal." In the end, it little mattered. Given the opposition of Train and others, no environmental reorganization occurred.[31]

Train maintained a few official connections to the Carter administration and met with the President several times. As Senator Jim Eastland noted Train "the kind of Republican the Administration should have kept," Carter reappointed Train as the American representative to NATO's Committee on the Challenges of Modern Society (CCMS). Periodically helping others just as they had helped him, Train recommended several administration appointments.[32] For the most part, however, he remained silent about the administration in public, saying little for the record as Carter struggled in environmental policy. Initially opposing large-scale water resources development, Carter angered western interests and eventually capitulated in dam appropriations. Despite executive orders protecting wetlands and flood plains, Carter surrendered his opposition to the Tennessee Valley Authority's Tellico Dam, a compromise that threatened at least one endangered species. To alleviate energy shortages, he proposed an Energy Mobilization Board with the ability to exempt power plants from environmental regulations. Through it all, unlike his environmental colleagues, Train offered no hard criticism.[33]

Train played no official role in Carter's major successes. When Carter signed the strip mining bill that Ford had twice vetoed, Train attended the White House ceremony. As Carter shook hands afterwards, Train "planted myself in [Carter's] path." The two talked for several minutes, Carter graciously saying it was "an honor" to have Train there. While Train's long efforts on the strip mining bill were obvious, he remained oddly quiet in regard to the two most important accomplishments of the Carter years, both signed late in 1980. The Alaska National Interest Lands Conservation Act placed 104 million acres in national parks or wildlife refuges. The Comprehensive Environmental Response, Compensation and Liability Act, commonly known as the Superfund Act, created a $1.6 billion fund for the cleanup of toxic waste sites and oil spills, financed by taxes on the petroleum and chemical industries. Train undoubtedly applauded both bills but did so quietly in the height of the political season.[34] Most noticeably, Train offered little praise when Carter launched a major initiative to curb the loss of the world's tropical forests, the concern that had prompted Train's own trip to the

Amazon. This program grew from a joint CEQ and State Department study pub-lished as the *Global 2000 Report,* an attempt to project environmental problems at the dawn of the new millennium. CEQ chairman Speth called Train to chair a citizen's committee on the report but Train declined. The 1980 election loomed.[35]

Bipartisanship was simply not easy in a partisan season. Train well knew that his friend George Bush wanted to be President. As early as 1978 Bush had asked Train for his support. Then visiting the Bushes at their Jupiter Island home not far from his own, Train enjoyed a cocktail party that included the entire Bush clan. "The young Bushes were all there," Train wrote, "George [W.] and his new wife Laura, Neil, Marvin, and Dorothy." Train assured Bush of his support but declined to "become publicly associated with an individual candidacy." Given his WWF responsibilities, it might cause problems. Barbara Bush encouraged her husband's candidacy but admitted that "when I stop to think, it terrifies me." Both Bushes assumed that former California governor Ronald Reagan could win the Republican nomination but would decline because of his age.[36]

Bush formally announced his candidacy on May 1, 1979. "I wanted to share that news with you in advance," Bush wrote Train a week before. "I am totally committed to making 1980 the Year of the Republican." Train limited his public partisanship to serving as a Bush alternate delegate at the Republican National Convention in Detroit. Privately, however, he committed to raising over $25,000 dollars for his friend. Campaign money was still tight, nevertheless, as Bush acknowledged in a form letter a year later. "I hate to ask, but could you send me $1,000, raised from a friend, a child of eighteen or older, or from anyone inter-ested? To get through these remaining primaries we need about 1.5 million dol-lars."[37] Working to raise money, Train quietly corresponded with colleagues who encouraged the strongest possible environmental stance.[38] Bush responded with forceful environmental rhetoric. "We no longer accept as inevitable the belch-ing smokestacks or polluted rivers of our past," his campaign declared. Criticiz-ing Carter's compromises, Bush urged stronger protections against toxic threats and acid rain, a better-organized and -staffed environmental bureaucracy, and—in what Train certainly appreciated—a renewed commitment to "our global ecology."[39]

Bush needed all the help he could get. Buoyed by social conservatives and exhibiting an optimism that the beleaguered nation demanded, Reagan took command in the primaries and won the nomination on the first ballot. Char-acterizing supply-side economics as "voodoo economics," Bush angered many conservatives who viewed him as the last remnant of "the East Coast wing of the party." His distinguished career in government won few delegates, who instead savored Reagan's attacks on liberalism and the federal establishment. Ironically, given his courage in World War II and his directorship of the Central Intelligence Agency, Bush suffered from the "wimp factor," a sense that he lacked the certitude that Reagan projected. The Republican platform reflected Reagan's influence, despite pleas for moderation from several environmental groups. Train read the platform critically but declined to testify with other environmentalists or issue a joint statement before the platform committee. "The World Wildlife Fund cannot join in any such statement," Train wrote, "and I have to ask that our name be removed." For her part, Aileen complained to Barbara Bush about the platform's strong antiabortion stance. The Reagans were fine people, Barbara replied, and there will always be issues of disagreement.[40]

The convention came with a surprise. After Bush had conceded defeat and gone to bed, Reagan selected him as his vice presidential nominee, an obvious attempt to appease moderates and unite the party. "You have done a fantastic job and I am proud to have had only a minor association," Train congratulated Bush. "Our hearts are with you and Barbara." Bush's selection encouraged Train, as did an invitation to serve on the Reagan-Bush Environmental Task Force. Train harbored doubts about Reagan's commitment to the environment but agreed to contribute a section on global environmental problems. Train initially did not know many of the group, writing in his diary that Stanford University's Hoover Institute "appears to be an intellectual center for the new right of the Reagan GOP." Nevertheless, he assumed his participation might mitigate the dominant influence of the conservatives. "If Reagan should be defeated, we certainly will not have to meet again."[41]

On November 4, Train voted for Reagan and Bush, "although with little enthusiasm for the top of the ticket." Party loyalty, his friendship with Bush, and hopes that he might influence the administration's environmental agenda proved pivotal. Train did not find Reagan's sweeping victory surprising given Carter's troubles, which included over a year of hostages held in Iran. The Republican control of the Senate, however, shocked many observers and launched what Train later described as a "lively time socially for Aileen and me." The Reagan task force "succeeded in charting a pretty moderate course," Train soon reported, "but we have no idea what the new Administration will do with our recommendations, if anything." Train arranged lunch with Bush and announced that he was not interested in any administration job, perhaps nervous of the direction the White House might take. The statement was rather presumptuous but Train "had a feeling that a lot of people who knew George would be seeking jobs and thought it well to eliminate the subject." Train, once again, knew how Washington worked. Several days before the inauguration, he invited the Bushes and an impressive array of Republican elite to his house for a black tie dinner. With furniture removed, there was music and dancing in the dining room. Included were reporter Diane Sawyer, Justice Potter Stewart, and Reagan's choices to lead the Commerce and Defense departments, Mac Baldridge and Frank Carlucci, respectively. Four secret service agents were in the house, a sign that the times had changed.[42]

Train had met Reagan several times, the first in Reagan's gubernatorial office when Train was EPA administrator. They had crossed paths at two Republican dinners but neither knew the other well. Another opportunity arose for Aileen, however, in the inaugural celebrations. Time, Inc. hosted a dinner for the President-elect at Washington's Renwick Gallery. There Aileen found herself seated between Reagan and *Time* magazine editor-in-chief Henry Grunwald. Aileen found Reagan a "very charming man," younger-looking and taller than expected. Grunwald asked Reagan if he worried over the task he faced. Apparently surprised at the question, Reagan smiled and replied that he was looking forward to it. Aileen tried to engage Reagan in a discussion of drug treatment but found him unresponsive. "He was not all that brilliant," Aileen remembered, Reagan more interested in talking about his movie career than substantive issues. Reagan at least listened; a week later Barbara Bush introduced Aileen to Nancy Reagan. Nancy replied that she had heard of the discussion and wanted to "be in touch when things settle down." Although the two never got together again, Aileen at least could claim to have encouraged Nancy's well-publicized antidrug campaign.[43]

On the day of the inauguration, the Trains went to Washington's Jefferson Hotel, where three buses took them and other Bush family and friends to the official ceremonies. "There was an immense crowd but quiet, orderly and good-humored," Train recalled. "Limousines were everywhere." Afterwards, a Georgetown restaurant catered a reception at the White House. There Train mingled with new administration officials and talked with famous evangelist Billy Graham. Employing his connections, Train won coveted seats outside the White House to watch the day's parade. At night, exhausted, the Trains decided to skip the ball. "It had been a great day but we were ready for bed."[44]

The new day brought a new reality. If Train hoped that the Reagan-Bush Environmental Task Force might influence the new administration, he quickly learned otherwise. Only a month after the inauguration, Train quietly bemoaned, "I am beginning to feel that everything I have done in the past is now going to be undone." Reagan's environmental appointments, for one, shocked the environmental community. In some respects Reagan tried to unite the moderate and conservative elements of his party. Encouraged by pragmatists such as White House aide Michael Deaver, Reagan appointed several moderates to his cabinet, including Labor Secretary William Brock and Education Secretary Terrel Bell. When it came to environmental policy, however, no doubt existed that conservative ideologues dominated.[45]

For Interior Secretary, Reagan selected James Watt. Not since Watt had outraged Senator Edmund Muskie decades before had Washington heard from the doctrinaire antienvironmentalist. In the years since, Watt had led the Mountain States Legal Foundation, a champion of business in environmental litigation. He had come to symbolize the "Sagebrush Rebellion," the drive by western ranching and mining interests to transfer federal land to the states. Upon learning of the nomination, Train called Watt, characteristically hoping to keep lines of communication open. "Jim," Train began, "it may not come as a surprise to you that you were not my first choice for the job." Watt agreed to attend a WWF board meeting, where Train had the opportunity to "argue quietly with him at the table." Little confrontation occurred, although "Walter [Cronkite] went after him on wilderness."[46]

Cronkite was not alone. After the environmental community unsuccessfully protested his nomination, Watt grew embittered—and more determined to fight. In one White House reception, Train approached him as he stood alone in the Red Room. "His eyes seemed to bulge at you," Train remarked, the man "very drawn and tense." Aileen thought Watt near a nervous breakdown. "Simple greed" motivated the environmentalists, Watt stated, their attacks meant to attract new membership. "It was plain," Train recalled, "that the situation had become further polarized."[47] Attempting to bridge the gap, Train later met Watt at the Department of Interior to present WWF's own Global 2000 Report. This study, a response to the Carter-era CEQ report of the same name, repeated many of the same projections and also called for action. Train said he was not there to add more drama but that it was important that the government have the capability to gather and analyze global resource trends. Watt replied that he had already heard of the report from MacGeorge Bundy. "When Mac said I should get into Global 2000 issues," Watt quoted himself, "I said, 'Mac, that's what we all are against.'" In later years scholars claimed Watt was inaccessible to environmentalists. This, however, was not completely true. Watt met the Republican Train; he just did not listen to him.[48]

EPA Administrator Ann Gorsuch was no better. A political protégée of Attorney General Edwin Meese, Gorsuch was a lawyer for a telephone company. She had briefly served in the Colorado legislature, representing the same interests that Watt supported, and had distinguished herself as a strict conservative. She had no experience managing a large organization, however, let alone a federal agency with complex scientific and technical responsibilities. Her primary qualifications included ideological loyalty and powerful friends. She, in turn, supported Rita Lavelle as head of the multibillion-dollar Superfund program. Lavelle, a public relations officer for a major industrial polluter, was a woman with a similar ideology and lack of experience.[49]

Initially, Train did not know what to expect. Reports of Gorsuch's connections to Watt and her failure to call and introduce herself did not bode well. "Doesn't sound like good news," Train wrote in his diary. As the months passed, however, Train learned how bad it was. Gorsuch claimed her job was to support "industrial revitalization" and lighten the regulatory "overburden" on industry. As she became embroiled in charges of mismanagement and favoritism toward polluters, Train privately complained that she "was in over her head."[50]

If Train thought the problem was solely one of appointments, he did not understand its depth. Watt enjoyed Reagan's personal support, who warned him that a thick skin was necessary to "take some of the [environmental] abuse that will be necessary to accomplish our objectives." Even as the evidence against Gorsuch accumulated, Reagan declared criticism against her "nothing but allegations and accusations." She was "doing a fine job."[51] Only three weeks after the election, Reagan's budget director, David Stockman, issued a public "manifesto" that predicted a "Republican economic Dunkirk" unless the White House swiftly curtailed federal regulations. With Reagan's backing, Gorsuch relaxed lead regulations for gasoline, lifted a ban on the disposal of liquids into chemical waste landfills, and adopted unenforceable voluntary testing and self-certification for the chemical industry. She developed proposals to exempt 60 percent of all new chemicals from the mandates of the Toxic Substances Control Act and eliminated reporting requirements for hazardous waste facilities, leaving acid rain and other hazards essentially unregulated. EPA adopted only three of twelve "new source reviews" required under the Clean Air Act and slashed its enforcement in almost every area. The number of EPA civil cases dropped by over 50 percent as Gorsuch personally reassured industry leaders. Watt, meanwhile, reduced parkland acquisition, diverted Land and Water Conservation Fund revenue to deficit reduction, and vastly expanded mineral leasing on public land.[52]

Deregulation was only one of Reagan's tools. Under the White House's regulatory reform plan, the Office of Management and Budget (OMB) had stronger control over EPA's purse-strings, which, in the words of historian Jeffrey Stine, was an "administrative rather than legislative tack."[53] OMB used the budget as leverage to eliminate or weaken restraints that hampered the conservative agenda. With its enforcement budget cut by over a quarter in Reagan's first two years, EPA reduced its staff by 11 percent. Budget cuts completely emasculated CEQ, which hardly had funds to carry out innovative studies and publish its annual reports. When rumors began that Reagan planned to eliminate the council, Train wrote Reagan's top aide James Baker to "urge that responsible members of the Administration find an early opportunity to sit down with CEQ's top staff and explore other options with them." The budget cuts hit close to home for Train. Administration plans not to fund CITES, he grumbled, were "outrageous."[54]

Unfortunately, however, Train remained largely quiet. In an uncomfortable position as his own party drifted away from his life's work, Train contented himself with private complaints. When WWF's board considered a resolution condemning the administration, Train strenuously argued against it. "I personally believe that while we should take up the cudgels on particular issues as we have always done when they are of direct concern to us," Train stated, "we should avoid adopting a position of confrontation." In one instance, Train even appeared indifferent toward Watt. "Some are trying to get Watt's scalp, but that is outside the traditional way the World Wildlife Fund operates. We tend not to be militant political activists." Train continued to repeat the same mantra as his colleagues angrily took up arms; WWF should "keep its lines of communication open" in hopes of moderation.[55]

Train had trouble realizing that, in this instance at least, moderation would not come from private pleas but public backlash. Train continued to lobby Bush even as Bush assumed the task of leading Reagan's regulatory reform. "EPA is a disaster," Train pleaded early in 1982. "The cuts in money and personnel go far beyond any effort to eliminate waste and inefficiency." They would, he insisted, "have an adverse political reaction." Surprising Bush, Train remarked that if he were head of EPA and faced such pressures, he would resign. When Train later repeated his complaints, White House aide and fellow moderate Dick Darmon agreed and even promised that all future policy decisions "would go through Bush's office with Boyden [Gray, Bush advisor] carrying the laboring oar." Train understandably doubted this and Darmon acknowledged that "it is almost impossible to get people to change their minds." It was, after all, more than simply agency decisions; it was the entire dominant conservative ideology.[56]

Train could take it no more, finally going public with a long article in the editorial pages of the *Washington Post*. "The Environmental Protection Agency is rapidly being destroyed as an effective institution," Train began. Citing the budget and personnel cuts, Train correctly noted that the agency was unable to fulfill its statutory obligations. Although a Republican, Train concluded that he thought EPA a "paper tiger." In many respects it was a harsh article—and, for the first time, very public criticism. Predictably, it drew an equally harsh reaction. Gorsuch called Train and invited him for a discussion, their first meeting. Train argued that she had become isolated from her career staff and should deal more with her "line people." Gorsuch defensively replied that she thought Train's complaints "had a political tone." Failing to reach any agreement, Train recorded his frustrations in his diary. "She doesn't seem to 'hear' criticism but simply tells you why it's not true." The meeting remained relatively amicable, "but I really doubt that I made much headway with her."[57]

Train underestimated his impact. Going public did produce results. Only four days after the article was published, Gorsuch distributed an agency-wide memorandum declaring the end of forced personnel terminations. She became more forceful against OMB's mandates and, by many accounts, appeared in retreat. It was not enough, however, to save her. Both she and Watt met the same fate as the environmental community applauded. With over a dozen top EPA employees already removed and with Lavelle under indictment, Congress investigated Gorsuch for illegally manipulating Superfund money. In the midst of the controversy, Gorsuch resigned. A year later the House Energy and Commerce Oversight Subcommittee concluded, "Top-level officials of the EPA violated their public trust by disregarding the public health and the environment, manipulating the Superfund program for political purposes, engaging in unethical conduct, and participating in other abuses." Watt's departure was less dramatic, his propensity for outlandish comments ultimately costing him his job. "A black, a woman, two Jews and a cripple" composed an advisory council, Watt declared, "and they have talent."[58]

The Gorsuch and Watt debacles attracted the media to the administration's environmental onslaught, drawing attention the environmental community needed. Political pressure began to build and, amazingly, the White House began its own retreat. To replace Gorsuch, Reagan selected William Ruckelshaus, the agency's first administrator. Reagan, Ruckelshaus recalled, "was shocked by the public reaction and began to back away." Train was quick with advice. The administrator, he told his longtime friend, should attend cabinet meetings and "have a free hand in personnel." He should "have the opportunity to review the entire EPA budget." It was a "high risk enterprise," especially since the Democrats "see EPA as their best political issue against Reagan." Ruckelshaus acknowledged the problems but replied that "the need is so great it will be difficult not to try."[59]

In the years that followed, Ruckelshaus helped mitigate administration policy. His greatest accomplishment was restoring to EPA professionalism and the respect of the scientific community. Ruckelshaus stressed the difference between risk assessment, a matter of science, and risk management, a matter of politics. Assuring that objective science was free from politics was an accomplishment in itself, but increased budgets soon followed. Risk assessment, in short, strengthened EPA's hand in dealing with OMB. As the White House backed away from staunch deregulation, EPA enforcement increased. Ruckelshaus remained at EPA only briefly but his successor, Lewis Thomas, helped strengthen the Clean Air Act and prioritized recycling, among other accomplishments. At Interior, William Clark, Watt's replacement, met with Train. The two got along well and Clark promised to appoint a liaison to environmental groups. Returning to an old refrain, Train described the "conservatism of many conservationists." Clark, like Ruckelshaus, restored credibility to his office. The administration pulled back from its efforts to convey public lands to the states and settled many lawsuits out of court.[60]

Although now undoubtedly aware of the power of public agitation, Train refused the media any further criticism of the administration and declined to testify at five congressional hearings. His *Washington Post* article remained an anomaly and, during the rest of the Gorsuch-Watt drama, he refused comment. When one journalist pressed for his opinion, Train recalled, "I got up and walked out of the room." Connections to the administration remained important and Train valued his Republican credentials. In one ABC television interview, he refused to join fellow guest Michael McCloskey of the Sierra Club in denouncing Reagan. "I wanted to talk about the future," Train remarked. The administration "should be open and fair to all, business as well as environmentalists." When Ruckelshaus faced criticism from the World Resources Institute, Train defended him.[61]

This was, ultimately, unfortunate. While the administration had moderated its environmental policies, it was hardly green. Even with Ruckelshaus's return, Reagan's two terms in office still represented the worst environmental record of the modern era. As Ruckelshaus acknowledged, "There were still many ideologues in the White House adamantly opposed to government regulation to affect social policy." In the words of one historian, Ruckelshaus's superiors "continued their posture of budgetary stringency toward the agency, as well as a generally anti-environment stance in policy issues."[62] The White House had shown itself to be sensitive to public reaction and fanning the flames might have engendered further concessions. For years Train's behind-the-scenes lobbying and his efforts at compromise had produced results. Now, however, he might have used his hard-won cache in a more effective manner. It was not Train's finest hour.

Budgets had improved but, in real dollars, had not returned to the levels of the previous administrations. The administration defunded one of Train's accomplishments, the World Heritage Trust, prompting from Train private—but not public—criticism. His pleas for a "fresh look at funding" fell on deaf ears. When Train contacted the administration about the possibility of reviving the U.S.-Soviet environmental accord, now ignored and largely moribund, he received no reply. While Ruckelshaus increased EPA enforcement, later years brought environmental complaints and lawsuits. The White House stoutly resisted Ruckelshaus's efforts to control acid rain and congressional efforts to expand research into indoor air pollution. It rejected proposals to protect parks from the development of adjacent lands and increased the timber harvest. Its proposals for wilderness additions were inadequate. Reagan's personal attitudes had not shifted; he had just realized that environmentalism was not so easily eradicated. As Reagan's first term neared its conclusion, Train and other selected environmentalists arranged a luncheon meeting with the President. It was, in Train's words, "a kick in the teeth." With no agenda, Reagan allowed no substantive debate. Only later Train learned that simultaneously Reagan had appointed Gorsuch, late of EPA, as chairwoman of a Department of Commerce air quality advisory commission. Unwittingly, Train and his colleagues had provided Reagan political cover.[63]

It was not the only time. In another instance, Train used his connections to have Reagan present the J. Paul Getty Wildlife Conservation Prize in a Rose Garden ceremony. "I don't need to welcome Russell Train of the World Wildlife Fund," Reagan joked. "Russell's more at home here than I am." After Reagan presented the awards, Train pinned a panda button on his lapel and declared him a member of the World Wildlife Fund. Several environmentalists complained but Reagan had his media moment. Bush's overtures were more genuine, meanwhile, although less politicized. Bush agreed to host WWF's board at the vice presidential mansion on the grounds of the U.S. Naval Observatory. During WWF's biannual meeting, Bush hosted hundreds of delegates for a barbeque under a large tent on the same grounds. "George and Barbara both have a knack for making everyone feel that this is without question the most enjoyable occasion of the year for them," Train recalled.[64]

For Train, Bush's kindness was representative of the old Washington. Somehow, however, those days seemed to have passed. Visiting the Kennedy Center to watch the ballet, Train heard an "undercurrent of booing" when the announcer introduced Reagan. "That's the first time anything of that sort has happened here," Train commented. "I have never heard Nixon or Carter booed in Washington."[65]

For every action, the old saying goes, there is reaction. Reagan's antienvironmental agenda "energized" the environmental community and much of the public, according to historian Hal Rothman. In some ways, radicalism begat radicalism. New environmental groups such as Earth First! formed, dedicated not to the incremental, compromising ways of Train but to confrontation—and, in some cases, violence. They practiced "ecotage," an amalgam of the words *ecological* and *sabotage*. Burning cars and spiking trees with deadly pieces of metal further polarized the debate. Emotions ran high on both sides. Most environmentalists thankfully avoided such atrocious tactics and found their numbers growing. Membership in such organizations as the Sierra Club and the Wilderness Society skyrocketed even as conservative resistance grew. In some ways the battle was more bitter. In the very least, however, Reagan had tested environmentalism and found it neither momentary nor superficial. In the words of historian Samuel Hays, Reagan "demonstrated the degree to which [environmentalism] had become a broad and fundamental aspect of American public life."[66]

WWF benefited from the backlash, the irony not lost upon Train in the age of Gorsuch and Watt. Under Train's auspices, WWF funneled public concern into environmental trusts, a unique mechanism for financing areas of particular natural interest. Train's first endeavor, the Bhutan Trust Fund for Environmental Conservation, began with a $1 million WWF grant. Other trusts used private funds and assistance from European and Asian governments. A number of environmental organizations adopted the concept, although almost half of the forty environmental trusts worldwide today have connections to WWF. Train also encouraged "debt-for-nature swaps," the idea originally outlined in the *New York Times* by Train subordinate Tom Lovejoy. Working with the World Bank and other organizations, Train recognized the debt crisis of many poor—and ecologically significant—countries as an opportunity. Foundations, government organizations, and various financial interests could purchase the country's dollar-denominated debt at the current market price, usually heavily discounted. They, in turn, could cancel the face amount of the debt and transfer funds equal to this amount to a private environmental endowment. The concept worked brilliantly and spread rapidly. In total, environmental trusts and debt-for-nature swaps raised over $1 billion for the natural world, Train a key player throughout.[67]

To help guide WWF, Train formed the WWF National Council, an advisory group that met twice a year, usually in conjunction with the regular board meeting. Train cultivated "warm friendships" with every member, ensuring effective and collegial management. He arranged for WWF to take over the African Fund for Endangered Wildlife, "lock, stock, and barrel." Most notably, he engineered a merger with his old organization, the Conservation Foundation. For years Train had lunched regularly with William Reilly, an early CEQ employee and friend, and then CF's president. At one such lunch, Reilly explained that he had declined an offer to lead the National Audubon Society. Their many chapters "were a headache," Reilly explained. Recognizing that Reilly wanted a more international role, Train recalled, a "light went off in my brain." If CF would join WWF, Reilly might serve as president, allowing Train to focus on his chairmanship of the board. Reilly accepted, their respective boards agreed, and both soon developed a strong working relationship. "Bill was a little nervous about working with me because I was a big figure," Train later explained. "He definitely wanted to run the show." At sixty-five years old, Train had no problem with this; he could focus on the big picture and Reilly the daily operations.[68]

This hardly meant retirement. Now elected to WWF's international board, Train won membership on its executive committee and, in turn, its vice presidency. As such, he worked closely with Prince Philip, the Duke of Edinburgh and the international board's president. Traveling through red-jacketed guards, Train visited Buckingham Palace and Windsor Castle. The two men got along relatively well, Train describing Philip as "stiff but always trying to do the job right." In fact, however, tensions existed. In one speech, Philip praised the American WWF but added, "Even I have a bit of trouble with Russ Train from time to time." It was more than a clash of personalities. With its tremendous success during the Reagan years, the American WWF grew more independent from the international coordinating body. Incorporated under American law, WWF was responsible for its own funds and could not automatically turn over significant money to WWF–International, as other nations' affiliates regularly did. As its revenues and programs expanded worldwide, WWF became a challenge to its sister organizations. "We didn't try to compete," Train explained, but WWF's growth was difficult to ignore.[69]

Success demanded that Train move carefully. Environmental groups guarded their donors and projects and, as WWF grew, Train had to reassure colleagues that he harbored no evil intent. Train corresponded with the leaders of other organizations to ensure cooperation and that their efforts did not overlap. In one instance, Train contacted the National Audubon Society to complain about its increasing role abroad. "Let's discuss this," came the reply. "We intend to work harder at [international projects], but our primary focus is and will be national."[70] Train's overtures paid dividends, although in some respects he might have been more judicious. WWF, on occasion, worked with brutal, dictatorial regimes in Asia and Africa, its efforts helping wildlife but arguably supporting the undemocratic governments. Despite pleas to the contrary, Train argued, for example, that WWF should remain in South Africa even as much of the western world boycotted the country because of its apartheid policies.[71] "We always try to be sensitive to the fact that we are working with cultures very different from our own," Train reported, rationalizing that boycotting the regimes meant abandoning the most ecologically threatened areas. "I am always hesitant to comment on political and social matters."[72]

Freed from much of WWF's daily operations, Train traveled more. In one trip to China to save the endangered panda—WWF's symbol—Train visited a 6,000-year-old Neolithic village, complete with a Confucian temple and large stone sculptures of lions, tigers, and rhinos. In a trip to Thailand arranged by the American ambassador, William Brown, Train and Aileen met King Phumiphon. It was important, Brown explained, that no one show the king the soles of his feet, a serious breach of protocol. When servants brought tea, they did so crawling on their knees. "I was talking to a general and all of a sudden I looked around and he wasn't there," Aileen recalled. "He was kneeling on the floor because the king had come over to talk to him." The Trains' travels augmented their long list of famous associates. After walking for hours on the Tibetan plateau, Train met Sir Edmund Hillary. The conqueror of Mount Everest remained "friendly, quiet and modest despite being an heroic figure in Nepal." On another occasion, Train met Jacques Cousteau and his son, Jean-Michael, champions of the oceans. Train kept in close contact with renowned naturalist Jane Goodall, who wrote that it was good to visit England and take time off from her African duties. In England she could enjoy ice cream. Closer to home, Train met famous actors Gregory Peck, Zsa Zsa Gabor, and Jimmy Stewart. Stewart was "old and somewhat feeble" but Gabor "caused trouble as usual." It appeared that almost everyone liked Russell Train, who was always polite and diplomatic.[73]

The Reagan years slowly passed with Train entering the age at which many people retreat to the golf course and rest on their laurels. For a man who liked to camp and hike around the world, however, Train was not yet ready for the gold watch and rocking chair. When the tenth anniversary of the U.N. Conference on the Human Environment approached, the famous Stockholm Conference in which he had played such a key role, Train agreed to travel to Nairobi for the celebration. There Japan recommended creation of the World Commission on Environment and Development to establish long-term strategies for achieving "sustainable development." The Reagan appointees resisted the commission but Train convinced them otherwise. For years, Train argued, America had encouraged Japanese involvement in the international environment. Denying them now was hypocritical. With the support of the American ambassador to Kenya, Train's arguments carried the day. The Brundtland Commission, named for its chair, Norwegian Prime Minister Gro Harlem Brundtland, defined sustainable development as allowing for the needs of the present without compromising those of the future. When the Brundtland Commission issued its report, *Our Common Future,* the Reagan administration ignored it but the definition of sustainable development remained as an accepted objective for the international community.[74]

Known for his mediation, Train agreed to serve on the Secretary of Defense's Commission on Base Closure and Realignment, a task as acrimonious as the Storm King debate. Outdated and unnecessary military posts dotted the land but remained important to their local economies. Congress had wrestled with the issue for years and tensions frequently ran high. Train visited several bases, met with his fellow commissioners, and hammered out compromises. Train's presence ensured that each closure met environmental as well as economic criteria. He was, after all, no stranger to controversy. The conclusion was "gratifying," he wrote a fellow commissioner. "I think we can take a good deal of satisfaction with the product of our effort."[75]

Retirement was simply not an option for Train until the Reagan years had passed. Fortunately, in Train's view, George Bush was Reagan's obvious Republican successor. Aware of Bush's historic moderation and his generosity to WWF, Train agreed to serve on the Bush '88 Citizen's Service Coalition. Members promised to "reach out to particular groups to tell about how the Vice President has encouraged and promoted citizen service, voluntarism, and private sector initiative." More importantly, Train agreed to join Ruckelshaus in cochairing the Conservationists for Bush Committee. Eight years before, Train had declined such an active role in his friend's campaign. Now, however, the situation was different. More settled at WWF, Train undoubtedly wanted to ensure that no new archconservative emerged. Perhaps, after all, he now recognized the need for more public activism.[76]

The job was not easy. Environmentalists flocked to Bush's Democratic opponent, Massachusetts governor Michael Dukakis. Bush's eight years in the Reagan administration, they feared, did not bode well. "We went through Hell and high water," a fellow committee member wrote Train, to win supporters. Few national environmental leaders joined, although the list included a large number of local environmentalists and prominent individuals such as Teresa Heinz, wife of Republican Senator John Heinz and future spouse of Democratic presidential candidate John Kerry. Meeting with New Jersey governor Tom Kean, pollster Bob Teeter, and other Bush supporters, Train helped craft a strong environmental message. When the League of Conservation Voters gave Bush a "D" on its environmental report card, Train and his colleagues correctly argued that the grade was a "misrepresentation." It "represented the Reagan Administration's appointments and actions" and not Bush's. Bush, Train claimed, had a strong congressional record and had helped mitigate Reagan's excesses. In a move that Train most likely resisted, the committee endorsed a controversial television advertisement that tied Dukakis to the pollution of Boston harbor.[77]

Bush reflected Train's influence, promising a "conservation ethic," a stance that the Washington Post declared "puts [Bush] at odds with the Reagan administration policy." Bush advocated protecting wetlands and strengthening both the Clean Air Act and Superfund laws. Train made one surrogate appearance for Bush in Ohio and traveled with him for an environmental speech in Michigan. On this latter trip, Bush asked Train to explain the ozone problem. An embarrassed Train "flunked" the answer but rebounded days later with a detailed letter. As the Republican National Convention approached, Train testified before its platform committee. At the convention, Bush once again distanced himself from Reagan in stressing the environment.[78]

Bush's victory in the general election came as a great relief to Train, although obviously many Republican conservatives harbored doubts. Writing his old friend, who was soon to become the most powerful person in the world, Train encouraged strong action to follow a strong campaign. Aware that Bush shared his international interests, Train recommended the environment as a "priority element of U.S. foreign policy." Writing in the Conservation Foundation newsletter, Train argued that the election gave Bush "a mandate" for environmental protection. "It must now be obvious from statements in the campaign itself, from voting patterns, and from opinion polls," he declared. Perhaps forgetting for a moment the power of the conservatives in his own party and the partisanship that blanketed the entire issue, Train displayed an optimism on national policy and politics that had been in hibernation for almost a decade. Environmentalism had survived the initial Nixon and Ford retreats and had proved resilient during the long Reagan attack. Now, with the moderate Bush in charge, a man with whom he shared much, a new dawn might be at hand. Washington might return to the days of old and the Republicans reclaim the banner of environmentalism. Then he could retire as confident as possible for the generations to come.[79]

EPILOGUE

Retirement?

In extolling the virtues of nature, the famous transcendentalist Ralph Waldo Emerson once wrote, "A man builds a fine house; and now he has a master, and a task for life; he is to furnish, watch, show it, and keep it in repair, the rest of his days." It was a sentiment that Russell Train could certainly appreciate— although not in the sense that Emerson intended. Train's house, after all, was not the structure of brick and wood that Emerson bemoaned. It was nature itself, Train's legacy of environmental law and organization. It too required constant vigilance lest it fall into disrepair.[1] In the winter of his life, Train took pride in his accomplishments. As one of the nation's most powerful environmentalists, he had helped protect for future generations their birthright of a clean, natural environment. His fingerprints lay on over a quarter-century of environmental legislation and enforcement. With his friend George Bush entering the Oval Office, the future appeared brighter. At the same time, however, Train sensed that the conservative forces of reaction were as resilient as American environmentalism. Even as retirement loomed, he could not escape one harsh reality: the battle was not won; indeed, total victory might forever prove elusive. Train's home—environmental protection—required a commitment that extended far beyond his own life.

Initially, at least, all seemed to be in order. Bush selected William Reilly as the new EPA administrator, an appointment that appeared to support the campaign's rhetoric. Pleased, Train advised his WWF colleague to keep a journal, exercise regularly, and develop a public image early so that opponents could not paint him unfairly. Reilly kept no journal but took the rest of Train's advice to heart. When Reilly's picture appeared on the cover of the *New York Times Magazine,* Train even warned Reilly that too much publicity risked alienating EPA subordinates. For his part, Bush fulfilled his promise to strengthen the Clean Air Act. The Clean Air Act Amendments of 1990 created a national market for sulfur oxide emissions. The law capped total emissions significantly lower than they had been a decade before but allowed industry to buy and sell pollution permits nationwide. In time, such emissions trading produced promising results. Backing Reilly in a number of initiatives, Bush allowed EPA to promote international research and technical cooperation on global warming.[2]

His pace slowed with age, and undoubtedly reticent about overly exploiting his relationship with the President, Train played little role in such developments. He convinced Bush to create a presidential medal recognizing environmental achievement, the President's Environment and Conservation Challenge Awards, and received several accolades himself. After Train once again helped frustrate efforts to abolish CEQ, chairman Mike Deland asked him to write the first chapter of the council's twentieth anniversary report.[3] For the most part, however, Train devoted himself to his WWF duties. He led efforts to establish a United States Committee for the World Heritage, run from his WWF offices. The purpose, Train told members such as the Sierra Club's Michael McCloskey, "is to demonstrate that the [World Heritage Trust and Convention] does in fact have a constituency in the U.S. and, thus, help sustain modest support for the concept." Largely free from the chronic health problems that plagued many of his contemporaries, Train still maintained an impressive travel schedule, including several trips to Africa that dovetailed nicely with his career. A trip to Antarctica that brought churning seas and threatening ice undoubtedly would have daunted many men half his age. In each instance, Train dutifully reported back to his WWF colleagues, essentially serving as a goodwill ambassador for the organization and the country.[4]

As he haggled with Prince Philip and WWF's international coordinating body, Train never veered from his philosophy of moderation and mediation. He agreed to serve as chairman of Clean Sites, Inc., an organization of environmentalists and industrialists dedicated to nonadversarial approaches to the contentious question of Superfund implementation and the cleanup of toxic wastes. He also agreed to serve on the board of the AES Corporation, an independent power company that operated a large number of electric power plants around the globe. Although in an industry known for its pollution, the company stressed social responsibility and maintained environmental standards much more rigid than its competition. The position seemed a perfect fit.[5]

With one eye on his legacy and the other on the future, Train led efforts to create a National Commission on the Environment (NCE). Unable to persuade the administration to participate, Train organized under WWF auspices an independent commission "diverse in composition and unfettered by political constraints." Composed of academics, private environmentalists, clergy, and representatives of business, NCE was a "bipartisan effort that brought together a broad spectrum of the interests involved in environmental issues." Meeting for over a year and a half in San Francisco, Washington, and Wisconsin, NCE neatly summarized the arguments that Train had made throughout his life: energy and environment did not have to be in conflict; environmental protection needed to consider the interests of man as well as nature; sustainable development was crucial; and success demanded cooperation and not political conflict.[6]

The latter goal, as Train's own life foretold, remained elusive. Much to Train's dismay and reflecting a similar trend in previous administrations, Bush found conservative resistance increasingly formidable. As Train watched rather helplessly, a familiar frustration grew; even his friend George Bush began to retreat from environmental protection. Bush's support for the new Clean Air Act amendments infuriated conservatives. "What was [Bush's] promise to be the environmental president," asked one, "other than an endorsement of the libel that Reagan had trashed the wilderness?"[7] Hearing such complaints, Bush at first refused to attend the 1992 U.N. Conference on Environment and Development in Rio de Janeiro, Brazil, and only belatedly agreed. This "Earth Summit," marking the twentieth anniversary of the Stockholm Conference, was the single largest gathering of world heads of state in history. Building upon the Brundtland Commission, it encouraged sustainable development and negotiated the Biological Diversity Convention. Delegates also reached an agreement to reduce carbon dioxide emissions by the turn of the millennium, thereby slowing climate change. Of the almost 8,000 representatives of NGOs present, Train was one of the few to have direct access to Bush. At one of the conference meetings, Train urged Bush to sign the biodiversity agreement but received no reply. In another meeting, Bush avoided such substantive discussions. "Train, how do you keep so fit?" he began. "Mr. President," Train deadpanned, "I eat anything I want, exercise as little as possible, and have two good, stiff martinis before dinner every night."[8]

In all fairness, the Earth Summit demanded unreasonable financial compensation from the United States and other developed nations. Nevertheless, Bush's reaction demonstrated that he, just as Ford and Carter before him, feared the power of the new conservatives. Alone among the developed countries, the United States refused to sign the Biodiversity Convention. It agreed to reduce greenhouse gas emissions only after defeating binding targets and deadlines. Train watched Bush's speech from his hotel room. Down in the lobby afterwards, Train met Tennessee Senator Al Gore and asked him what he thought of the President's speech. "Disgusting," Gore replied.[9]

Train was more dismayed than disgusted; he had seen it all before. Train had never been a radical. He agreed with the administration's advocacy of the North American Free Trade Agreement (NAFTA), which once again placed him at odds with many of the newer, more activist environmental groups, and supported Reilly's risk-based analysis at EPA. Nevertheless, Bush's sudden embrace of the New Right—as the conservative movement was now known—had an all-too-familiar ring. Bush wavered in his support of international family planning, increased clear-cutting in national forests, and appointed many federal judges whose narrow interpretation of the constitution's commerce clause inhibited much environmental legislation.[10] Despite his campaign promises, Bush overruled Reilly and changed the definition of wetlands. The more narrow interpretation allowed for additional development and generated over 80,000 complaints, including one from Train. The debate continued unresolved throughout Bush's term. Reilly slowly emerged as a pariah at the White House, compromising his ability to champion reform inside the Executive Branch. Vice President Dan Quayle's Council on Competitiveness counteracted extant environmental regulations and opposed new legislation. As one historian noted, "The [Quayle] council quickly became controversial because of its pro-industry bias and power to change regulatory decisions outside the view of the public and Congress." In a larger sense, Reagan-era budget deficits proved as much an obstacle to environmental protection as ideological bias did. Train remained quiet as he had in the past, leaving it to others to howl publicly. David Brower, an occasional Train critic, was never so restrained. "Thank George Bush for saying he is an environmentalist," Brower stated, "and thank him again when he becomes one."[11]

Brower's rhetoric was typical. While Train longed for the compromises and bipartisanship of the past, anger and confrontation was increasingly the rule of the day. "I thought the essence of good government was reconciling divergent views with compromises that served the country's interests," Senator Warren Rudman recalled. "But that's not how 'movement' conservatives or far-left liberals now operate." At Clear Sites, Train floundered despite his best efforts. He counted a few minor successes but understood that strong incentives existed not to cooperate. Businesses concluded that lawyers were cheaper than contractors, while environmental organizations determined that public outrage increased membership and contributions.[12]

Train continued in the ways of old Washington, inviting members of NCE and the Clean Sites' board to his home for dinner and drinks.[13] Around him, however, practitioners of a new viciousness waged political warfare. Democrats derided Bush's nomination for Secretary of Defense, John Tower, claiming that he had a drinking problem and was beholden to defense contractors. Republicans, led by the upstart young Georgia congressman Newt Gingrich, responded by toppling Speaker of the House Jim Wright, alleging financial improprieties. "When harsh personal attacks upon one another's motives and one another's character drown out the quiet logic of serious debate on important issues," Wright stated in resigning, "surely that is unworthy of our institution and unworthy of our American political process."[14]

The victims were manifold. Despite his attempts to appease them, Bush angered conservatives by raising taxes. Without strong conservative support, Bush lost reelection in 1992 to Arkansas governor Bill Clinton, who became the first two-term Democratic President since Franklin Roosevelt. The Clinton Presidency, in turn, resulted in the infamous $40 million Whitewater investigation and, ultimately, acquittal in an impeachment trial. In many respects, the partisan wars hardly benefited the environment. For many the environment had joined taxes and a litany of social concerns such as abortion and gay rights as wedge issues, defining one's partisan allegiance. When Texas Republican congressman Tom DeLay called EPA "the Gestapo of government," environmental support for Democrats increased. When environmental radicals torched large, gas-guzzling sports utility vehicles, Republican strength grew. More than ever before, the environment provoked divergent reactions among party faithful. Environmental protection required bipartisan agreement but, in the new era of attack politics, compromise had become a sign of personal weakness and ideological impurity.[15]

Train played no role in Bush's defeat, declining to chair an environmental campaign committee. Four years before, he had agreed to do so but now claimed that his role with NCE precluded it. Undoubtedly, however, Bush's new conservatism made the decision easier. At the same time, Train maintained little contact with the Clinton administration. When NCE completed its report late in 1992, *Choosing a Sustainable Future,* many members urged its release during the presidential campaign. Train thought this unwise in such a political climate and convinced his colleagues to delay. If Train hoped the new Clinton White

House might welcome the report, he learned otherwise. Despite rhetoric that far surpassed the Bush campaign's, Clinton largely delegated the environment to Vice President Al Gore. Gore, the author of his own book on the environment, offered environmentalists' hope but had his own ideas on the best path to take. "He ignored [NCE's] report," Train recalled. "He felt that if it wasn't coming from his own shop, it didn't matter."[16]

A decade and a half earlier, Train had watched environmentalists' hopes for the last Democratic President, Jimmy Carter, drown in a conservative tide. Now, with many environmentalists having dubbed Clinton the "Great Green Hope," it was eerily familiar. Like Carter, Clinton counted several successes. Train welcomed the new EPA administrator, Carol Browner, and the new Secretary of Interior, Bruce Babbitt, the latter an old acquaintance whom Train congratulated personally. Browner thought NCE's report impressive and disseminated it among her agency subordinates.[17] Pledging to make his administration "the greenest in history," Clinton signed the California Desert Protection Act, which added 7.5 million protected acres, and disbanded Quayle's troublesome Competitiveness Council. He reactivated the dormant Soviet-American environmental agreement, now with the Russian government, as Train applauded. With Train's support, Clinton proposed elevating EPA to the cabinet level and abandoned early plans to abolish CEQ in favor of a White House Office on Environmental Policy. Despite initial resistance, he ultimately accepted Train's arguments that any new body "would lack statutory authority and undermine implementation of the National Environmental Policy Act."

Although Train's son Bowdy lost his lower-level EPA job with the new regime, Train had little reason for complaint—at least in the first few years. Just as so often in the past, conservatives reasserted control. The 1994 midterm congressional elections looked, to many observers, like the ultimate victory of the New Right. Waving their "Contract with America," new conservative majorities sought to dismantle the federal bureaucracy in a way not seen since Reagan's first term. *Business Week* magazine spoke of "the GOP's guerrilla war on green laws." Clinton, who already described himself as a "New Democrat," reacted by moving to the political right. "As the GOP became more conservative," Michael McCloskey recalled, "the Democrats were busy trying to find a centrist approach not harsh to the business community."[18]

Train had seen it all before—Nixon, Ford, Carter, Bush, and now Clinton. In each instance hope and optimism had met conservative reality. Accomplishments were many but with nothing in stone. Environmentalism had persevered but the nation's body of environmental law and regulations still faced a political world more angry and partisan than Train and others of his generation had ever thought possible. "Neoconservatives," environmentalist Joe Browder reflected, "used the environment to paint Democrats as extremists." Liberals "used it to whip up support against the Republicans."[19] Moderation and bipartisanship appeared dead and, despite years of effort, the nation's environmental future still remained tenuous. Retirement, Train now undoubtedly recognized, might not bring the contentment for which he had hoped.

Two months before the Republicans swept to victory in the 1994 congressional elections, the long illustrious career of Russell Train came to a close. With his nine-year term as WWF chairman complete, the seventy-four-year-old Train stepped down in favor of Roger Sant, a fellow board member and Ford administration veteran. Predictably, many of Train's friends did not plan to let him go quietly into the good night; his retirement warranted a celebration worthy of his career. Kathryn Fuller, WWF president since Reilly's departure, organized an honorary dinner at Washington's Union Station. The refurbished and ornate turn-of-the-century train station near the Capitol was the perfect location, if an obvious play on Train's name. The building included a large hall with high ceilings, perfect for the over 250 guests in attendance. The crowd reflected Train's career—prominent politicians, environmentalists, and industrialists among the guests. Former Nixon administration officials such as Elliot Richardson joined journalists such as Rowland Evans. CEQ veterans traded Train jokes with current WWF staff. Neighbors mingled with the powerful, all united at least in their admiration for Train.

Walter Cronkite agreed to host the ceremony, which included a large screen for pictures of Train's life and accomplishments. After Fuller spoke, Train got up to thank the crowd. Cronkite, however, interjected, "Hey, it's my turn." Everyone laughed but the torrent of praise was genuine. Participants contributed letters of reminiscence, stories of Train's humor and good nature during the most stressful times. "The more I have thought of Russ's retirement, the more I have discovered my incapacity to think of Russ being out of wildlife and environmental matters," wrote one colleague. "The more I have thought of it, the more I have felt that we should quietly plot to refuse Russ a retirement, and fondly roll him onto the next logical assignment, i.e., that of being 'Ambassador Plenipotentiary for Wildlife and the Environment' for the rest of his life."[20]

Fuller announced the greatest honor—the Russell E. Train Education for Nature Fund (ENF) to provide scholarships, fellowships, and institutional grants for environmental education. Included were the Russell Train Conservation Leadership Awards for additional professional development. WWF had already raised $6 million toward the $10 million endowment goal, Fuller declared, Train's many admirers more than willing to match their applause with cash. "A generation of people in the field of conservation will learn not just from the skills and intelligence of Russ Train but also from his philosophy, integrity, and moral perspective," stated Reilly. There was a symmetry to ENF. Three decades earlier Train had launched his conservation career educating Africans; ending it in a similar vein seemed appropriate.[21]

"I am now chairman emeritus," Train wrote friends after the ceremony, "whatever that means." With a title, an office, and a secretary but almost no official duties, Train's view of retirement still included environmental advocacy. From his perspective, the state of the environment demanded it. Train certainly had more time for his family and love of reading, adding to his collection of over 3,000 books and artifacts on Africa. Eventually Train cataloged each item and donated the collection to the Smithsonian Institution. For years Train had dabbled in horse racing and now purchased shares in an English racing syndicate. Trips to the Royal Ascot followed, complete with top hat and traditional attire. Train also purchased the Bachelor Point Harbor Marina near his Eastern Shore farm. Still, however, environmental politics was never far away, if only in discussions with friends and former colleagues.[22]

Each day began with the newspaper—and the news, unfortunately, suggested that the past was only prologue. It was an old man's worst nightmare. The new Republican President George W. Bush, elected as Train entered his ninth decade in 2000, launched a frontal assault on the environment that surpassed even the Reagan era's. Repudiating the moderation that had defined his father and the pragmatism that had characterized Reagan's second term, the younger Bush seemed the culmination of the conservative ideology hostile to environmental protection. This ideology had blunted the environmentalism of his immediate two predecessors but now commanded the White House as never before. Unlike Reagan, the new Bush couched his plans in green rhetoric, a tactic that appeared to mute public opposition. He employed "subtle, legalistic ways of cranking back on the environment," noted Michael McCloskey, "things that were not easily understood." After the September 11, 2001, terrorist attacks, the policies continued but garnered almost no publicity, the media preoccupied with wars in Afghanistan and Iraq. The result was what the Sierra Club termed "crimes against nature." George W. Bush, one critic bluntly wrote, "is the worst environmental president in American history."[23]

Each new development represented a body blow to Train's lifework. Under what he termed his "Clean Skies" proposal, Bush circumvented many Clean Air Act caps that had limited emissions trading since his father's day. Reinterpreting the law without publicly rejecting or revising it, Bush made the monitoring of compliance more difficult. Similarly, without criticizing the law's mandated "new source reviews," provisions that required pollution control upgrades for older plants, he quietly adopted new rules that gutted the requirement. While praising the Clean Water Act, he cut funding and enforcement behind the scenes. A redefinition of wetlands dramatically narrowed protected areas. Under a plan termed the "Healthy Forest Initiative," Bush overstated the threat from wildfires to increase significantly logging in the national forests. With no major publicity, he opened federal lands to almost every extractive industry. When environmental groups sued to expose the influence of large corporations in the formation of the administration's energy policy, the White House fought successfully to keep the record out of the public eye. The assault was broad and audacious even as Bush professed environmentalism. Through it all, Bush enjoyed a compliant Republican Congress; publicity thwarted only his proposal to open the Arctic National Wildlife Refuge to oil drilling. Americans rallied to their commander-in-chief in an era that promised perpetual war, the natural world suffering but paling in comparison to the imminent threat of death and destruction from terrorist attack.[24]

Bush was the darling of most conservatives. Rejecting global warming as "hot air," columnist George Will defended the new administration. "There is no question that there is global warming," Train privately complained, "and there is no question that humans have made at least some contribution." Initially, Train thought the new administration inept. Once it gained experience, he hoped, Bush might retreat somewhat like Reagan before him. As time passed, however, Train began to believe that corporate polluters wielded excessive influence over the White House, which, in turn, let politics guide regulatory decisions. With his EPA experience, Train found this particularly annoying. "The White House is calling the shots at EPA," Train complained, "which was not the way we did business when the agency was set up." Bush selected former New Jersey governor Christine Todd Whitman as EPA administrator, a moderate Republican and a choice Train applauded. Bush, however, overruled her promises to add carbon dioxide to the list of regulated pollutants and it became apparent that the White House limited her authority. "I don't think she ever had the opportunity to do the job she would have liked to have done," Train concluded. A good soldier never complaining while in office, Whitman finally resigned and penned a book with more criticism.[25]

As his frustrations grew, Train kept telling all who would listen that "conservatives are the real conservationists." He kept reminding them that Teddy Roosevelt was a Republican and that Nixon had signed the nation's most important environmental legislation. It was true, of course, but appeared irrelevant given the political reality. Asked what he thought of the present, Train more often than not demurred. Asked for whom he was going to vote in the next presidential election, Train claimed the campaign too distant to speculate.[26]

When the 2004 election finally arrived, Train faced the decision of a lifetime. He had always been a loyal member of the Republican Party, building on its strong environmental record when possible and muting its retreat when necessary. He had voted for every Republican candidate since Thomas Dewey, even Reagan. Now, however, his party was not the same. The new conservatives were finally dominant, their antipathy toward the federal bureaucracy no longer restrained. The forces of deregulation had their champion in George W. Bush. At the same time, civility in public life had become so rare that no hope existed for the old Washington. Politics would forever remain brutal, the common good lost in the struggle for partisan advantage. It was all too much and required drastic action. With the election less than a year away, Train made his decision; he would vote for Massachusetts Senator John F. Kerry—the Democratic candidate.

Train was not alone, nor his decision made in private. Joining with a number of moderate Republicans, Train employed the political activism that he had long avoided. "At eighty-five and at a time when he could otherwise be savoring crisp fall days on his farm," the *Baltimore Sun* reported, "Russell E. Train is out on the campaign trail in battleground states around the country." Part of a group known as Republicans for Environmental Protection, Train stressed that the environment was "too important to swing on a pendulum every four years, attached to a political party or to a liberal or conservative label." He was a Republican, he maintained, but would vote to quash Bush. Giving more radio and television interviews than in any previous election, writing articles in numerous newspapers, and giving speech after speech, Train was a new man for a new time.

Always diplomatic, he was now blunt. Bush had declared "war on the environment." His policies were an "abomination." In short, everyone who cared about the environment should vote Democratic.[27]

The irony was obvious. Train was retired but was more active than ever; he was a champion of consensus but was on the political attack. Train had watched for decades the Washington he loved slowly change and the party he admired slowly evolve. Even in the sunset of his life, he would have to evolve as well. The task remained incomplete, his house—environmental protection—demanded it. For Russell Train, there was no real retirement.

Bush's narrow reelection on November 2, 2004, turning on terrorism and the culture wars, owed little to the struggle for environmental protection. It was, nevertheless, a critical election that appeared to repudiate both Train and his cause. In reality, Train's long record of environmentalism was not so easily dismissed, despite the melancholy of environmental advocates from coast to coast. History has yet to render its final judgment on the Bush administration, but no President can escape the reality of what Russell Train consistently preached: it is a small world in which our survival depends upon a harmony with nature.

In his memoirs, Train called for an "environmental ethic." We must, he wrote, "see ourselves, individually and collectively, as part of an interrelated, interdependent community: the community of living things, the world of nature."[28] It was a constant philosophy that guided Train's life even as the world around him continued to change. He had witnessed the nation turn from wise-use conservation to environmentalism; he had led the adoption of strong environmental laws only to combat a new, reactionary conservatism; and he had watched the political culture of Washington grow more coarse and the tenets of his own beloved Republican Party become more ideological and narrow. He had made mistakes in adjusting, with his moderation occasionally insufficient to meet the challenge. The result was far from perfect. Nevertheless, he knew that George W. Bush was only one opponent in an ongoing war. Train's long life was an amazing story in this enduring conflict, a man of the elite who combined his good fortune with hard work and determination on behalf of the common good. Today his legacy is evident from the savannahs of Africa to the boardrooms of international conglomerates, found not only in thriving organizations such as the World Wildlife Fund but in the minds and aspirations of thousands of Americans. The struggle for environmental protection will continue without him but, hopefully, with others who share his passion.

NOTES

INTRODUCTION

1. Letter, Scott McVay to Kathryn S. Fuller, June 1, 1991, Public Papers of Russell Train (hereafter PPRT), uncataloged, Manuscript Division, Library of Congress (hereafter LC).

2. The following institutions awarded honorary degrees: Bates College, Clarkson College of Technology, Columbia University, Drexel University, Michigan State University, Princeton University, Salem College, Saint Mary's College, University of Maryland, University of the South, and Worcester Polytechnic Institute.

3. Other awards included the Lindbergh Award, the Elizabeth Haub Prize in International Environmental Law, the John and Alice Tyler Ecology Award, the Frances K. Hutchinson Medal from the Garden Club of America, the Environmental Recognition Award from the Rene Dubos Center, the Environmental Law Institute Award, the Aldo Leopold Award from the Wildlife Society, the Albert Schweitzer Medal from the Animal Welfare Institute, and the Herschel C. Loveless Award for Environmental Achievement in the Public Sector from the National Environmental Development Association. Among others, Train served the following nonprofit organizations: African Wildlife Foundation; Alliance to Save Energy; American Conservation Association; Citizens for Ocean Law; Clean Sites, Inc.; Elizabeth Haub Foundation; King Mahendra Trust for Nature Conservation; Resources for the Future; Rockefeller Brothers Fund; Scientists' Institute for Public Information; World Resources Institute; and the World Wildlife Fund International. Train also served as a director of Union Carbide, Incorporated.

4. *Washington Post*, September 12, 1994, D3.

5. See J. Brooks Flippen, *Nixon and the Environment* (Albuquerque: University of New Mexico Press, 2000).

6. "A Tribute to Russell Train," Elliot Richardson, July 19, 1994, PPRT, LC.

7. Letter, Lynn Martin to Russell Train, December 3, 1991, PPRT, LC.

8. Diary of Russell Train, June 24, 1975, PPRT, LC.

9. Letter, Russell Train to George Bush, July 15, 1975, PPRT, LC.

10. Interview, Author with Russell Train, January 8, 2004.

11. Letter, Thomas Curnin to Russell Train, February 5, 1992, PPRT, LC.

12. Letter, John Hanks to Russell Train, November 15, 1991, PPRT, LC; Letter, Julia Chang Bloch to Russell Train, December 4, 1991, PPRT, LC; Letter, Siegfried Woldhek to Russell Train, November 18, 1991, PPRT, LC; Letter, J. Von Noordincjk to Russell Train, November 16, 1991, PPRT, LC; Letter, Charles de Haes to Russell Train, November 15, 1991, PPRT, LC; Letter, Tammenoms Bakker to Russell Train, February 11, 1992; Letter, D. Pilar to Russell Train, November 21, 1991, PPRT, LC.

13. Program, "On the Occasion of the Presentation of the Presidential Medal of Freedom, The White House, Monday, November 18, 1991," PPRT, LC; *New York Times*, November 19, 1991, 32.

14. Interview, Author with Russell Train, January 9, 2004.

15. Ibid.; Barbara Bush quoted in Diary of Russell Train, 1991 White House Visit, January 17, 1991, PPRT, LC.

16. Letter, Russell Train to Tammenoms Bakker, February 19, 1992, PPRT, LC; Diary of Russell Train, 1991 White House Visit, January 17, 1991, PPRT, LC.

17. Hal K. Rothman, *The Greening of a Nation?: Environmentalism in the United States since 1945* (Fort Worth, Tex.: Harcourt Brace, 1998), 16–18.

18. Samuel P. Hays, *Conservation and the Gospel of Efficiency: The Progressive Conservation Movement, 1890–1920* (Cambridge, Mass.: Harvard University Press, 1959), 267.

19. Richard N. L. Andrews, *Managing the Environment, Managing Ourselves: A History of American Environmental Policy* (New Haven, Conn.: Yale University Press, 1999), 201–226.

20. Interview, Author with Alice Rivlin, January 24, 2005; Diary of Russell Train, February 19, 1976, PPRT, LC.

21. Letter, J. D. Hodgson to Russell Train, January 9, 1973, PPRT, LC.

22. Interview, Author with George Hartzog, January 20, 2005.

23. Quoted in Katharine Graham, *Katharine Graham's Washington* (New York, N.Y.: Random House, 2002), 148.

24. Michael Schaller and George Rising, *The Republican Ascendancy: American Politics, 1968–2001* (Wheeling, Ill.: Harlan Davidson, 2002), 14–15.

25. Quoted in *U.S. News and World Report* (June 28, 2004): 76.

26. Interview, Author with John Whitaker, July 23, 1996.

27. Nixon quoted in Memorandum for the President's File, David Parker, March 29, 1974, Folder "Meetings File, Beginning March 24, 1974," Box 94, President's Office Files, White House Special Files (hereafter WHSF), Richard M. Nixon Presidential Materials Project (hereafter RNPMP), National Archives II, College Park, Md.

28. Interview, Author with Laurence Moss, March 25, 1998.

29. Quoted in *Environmental Quality Magazine* (June 1973): 19–26.

30. Interview, Author with Christian Herter, January 20, 2005; McCain quoted in *Dallas Morning News,* March 19, 2004, A9.

31. Samuel Hays, *Beauty, Health, and Permanence: Environmental Politics in the United States, 1955–1985* (New York, N.Y.: Cambridge University Press, 1987), 491.

32. Jeffrey K. Stine, "Natural Resources and Environmental Policy," in W. Elliot Brownlee and Hugh Davis Graham, eds., *The Reagan Presidency: Pragmatic Conservatism and Its Legacies* (Lawrence: University of Kansas Press, 2003), 235.

33. World Wildlife Fund–United States, *Annual Report, 2003* (Washington, D.C.: World Wildlife Fund, 2003), 34–35.

34. Interview, Author with Michael McCloskey, June 29, 2004.

35. Russell Train, *A Memoir* (Washington, D.C.: self-published, 2000), 293–294.

36. Ibid., 297.

CHAPTER 1

1. Letter, Thomas Jefferson to John Adams, October 28, 1813, in Merrill Peterson, ed., *Thomas Jefferson: Writings* (New York, N.Y.: Literary Classics of the United States, Inc., 1984), 1304–1310.

2. See Susan Train Hand, *John Trayne and Some of His Descendants* (New York, N.Y.: self-published, 1933); David Hackett Fischer, *Albion's Seed: Four British Folkways in America* (New York, N.Y.: Oxford University Press, 1989), 28.

3. Russell E. Train, *The Train Family* (Washington, D.C.: self-published, 2000), 5–7.

4. Quoted in ibid., 11–18.

5. Ibid., 22–23, 26–27, 46–47.

6. Ibid., 39–42.

7. Interview, Author with Russell Train, January 7, 2004.

8. The family referred to Charles Russell Train II as Russell, but in order to avoid confusion, this work will not.

9. See Alfred Thayer Mahan, *The Influence of Seapower on History* (Boston, Mass.: Little, Brown, 1890).

10. Edward L. Bench, *The United States Navy: 200 Years* (New York, N.Y.: Holt, 1986), 426–430; Train, *The Train Family,* 130–131.

11. Interview, Author with Russell Train, January 7, 2004.

12. Quoted in ibid.

13. Train, *The Train Family,* 151–160.

14. Interview, Author with Russell Train, January 7, 2004.

15. Quoted in Train, *The Train Family,* 170.

16. Quoted in ibid., 53.

17. Meg Greenfield, *Washington* (New York, N.Y.: Public Affairs, 2001), 24–25.

18. Russell E. Train, *Politics, Pollution, and Pandas* (Washington, D.C.: Island Press, 2003), 14; Robert Remini, *Andrew Jackson and the Course of American Freedom, 1822–1832,* vol. 2 (New York, N.Y.: Harper and Row, 1981), 169–170.

19. *Washington Post,* June 16, 1908, 21; Train, *The Train Family,* 151.

20. Bureau of the Census, *Historical Statistics of the United States: Colonial Times to 1970,* vol. 2 (Washington, D.C.: U.S. Government Printing Office, 1975), 640.

21. The family was at their summer residence in Rhode Island when Train was born; Interview, Author with Russell Train, January 7, 2004.

22. Ibid.

23. Russell Train, *A Memoir* (Washington, D.C.: self-published, 2000), 17.

24. Interview, Author with Russell Train, January 7, 2004.

25. *New York Times,* October 25, 1954, 31.

26. Interview, Author with Russell Train, January 7, 2004.

27. Train, *A Memoir,* 21–23, 38, 45–46.

28. Ibid., 66–67; Interview, Author with Russell Train, January 7, 2004.

29. Interview, Author with Russell Train, January 7, 2004.

30. David M. Kennedy, *Freedom from Fear: The American People in Depression and War, 1929–1945* (New York, N.Y.: Oxford University Press, 1999), 131.

31. Train, *A Memoir,* 62–63; see also Katharine Graham, *Personal History* (New York, N.Y.: Alfred A. Knopf, 1997), 103.

32. Train, *A Memoir,* 67; Interview, Author with Russell Train, January 7, 2004.

33. Interview, Author with Russell Train, January 7, 2004.

34. Ibid.; Train, *A Memoir,* 76–79.

35. Interview, Author with Train, January 7, 2004.

36. Handwritten Comments on Russell Train, Senior Thesis, "The United States versus Japan: A Study of Sea Power in the Pacific," PPRT, LC; Personal Data Statement, Russell Train, July 25, 1973, PPRT, LC.

37. Interview, Author with Russell Train, January 7, 2004; Train, *A Memoir,* 89.

38. Train, *A Memoir,* 104–106.

39. Interview, Author with Russell Train, January 7, 2004.

40. Train, *A Memoir,* 104–106.

41. Diary of Russell Train, undated, September, 1950, PPRT, LC.

42. Interview, Author with Russell Train, January 7, 2004; Train, *The Train Family,* 201.

43. Lay Sermon, Russell Train, "A Layman Looks at the Christian Faith Today," November 30, 1966, PPRT, LC.

44. Interview, Author with Russell Train, January 7, 2004.

45. See Donald R. Kennon and Rebecca M. Rogers, *The Committee on Ways and Means: A Bicentennial History, 1789–1989* (Washington, D.C.: U.S. Government Printing Office, 1989).

46. Ibid.; Biographical Summary, Russell Train, 1955, PPRT, LC.

47. See Elmo Richardson, *Dams, Parks, and Politics: Resource Development and Preservation in the Truman-Eisenhower Era* (Lexington: University of Kentucky Press, 1973); Mark Harvey, *A Symbol of Wilderness: Echo Park and the American Conservation Movement* (Albuquerque: University of New Mexico Press, 1994); and David R. Long, "Pipe Dreams, Hetch Hetchy, the Urban West, and the Hydraulic Society Revisited," *Journal of the American West* (July 1995): 61–84.

48. Interview, Author with Russell Train, January 7, 2004.

49. Letter, Howard Baker to Kathryn S. Fuller, June 22, 1994, PPRT, LC.

50. Letter, Elliot Richardson to Kathryn Fuller, July 19, 1994, PPRT, LC.

51. Quoted in Train, *A Memoir,* 115.

52. Train, *Pollution, Pandas, and Politics,* 24.

53. Interview, Author with Russell Train, January 7, 2004; Bureau of the Census, *Historical Statistics,* vol. 1, 289.

54. Transcript, Interview, Harold K. Steen, Forest History Society, with Russell Train, 1993, PPRT, LC.

55. Interview, Author with Russell Train, January 7, 2004.

56. *Miami Herald,* May 19, 1957, 18.

57. Eisenhower quoted in Biographical Summary, Russell Train, 1955, PPRT, LC; see Gary Reichard, *The Reaffirmation of Republicanism: Eisenhower and the Eighty-Third Congress* (Knoxville: University of Tennessee Press, 1975), and Charles C. Alexander, *Holding the Line: The Eisenhower Era, 1952–1961* (Bloomington: Indiana University Press, 1975).

58. Letter, Roswell Perkins to Leonard Hall, November 14, 1955, PPRT, LC.

59. Michael Schaller and George Rising, *The Republican Ascendancy: American Politics, 1968–2001* (Wheeling, Ill.: Harlan Davidson, 2002), 9, 19.

60. See Fred Greenstein, *The Hidden Hand Presidency: Eisenhower as Leader* (Baltimore, Md.: Johns Hopkins University Press, 1994).

61. Nixon quoted in Interview, Author with Russell Train, January 7, 2004.

CHAPTER 2

1. Interview, Author with Russell Train, January 7, 2004.

2. *Washington Evening Star,* January 28, 1972, C1.

3. Meg Greenfield, *Washington* (New York, N.Y.: Public Affairs, 2001), 117.

4. Interview, Author with Aileen Train, January 9, 2004.

5. Quoted in ibid.

6. See Russell Train, *The Bowdoin Family* (Washington, D.C.: self-published, 2000).

7. Interview, Author with Aileen Train, January 9, 2004.

8. Ibid.

9. Ibid.

10. Bureau of the Census, *Historical Statistics of the United States: Colonial Times to 1970*, vol. 2 (Washington, D.C.: U.S. Government Printing Office, 1975), 639.

11. Interview, Author with Aileen Train, January 9, 2004.

12. Interview, Author with Russell Train, January 7, 2004.

13. Letter, Russell Train to John Whitney, March 8, 1956, PPRT, LC.

14. Letter, Russell Train to Daniel Reed, June 15, 1956, PPRT, LC.

15. Ibid.

16. Letter, John Byrnes to George Humphrey, May 14, 1956, PPRT, LC.

17. Letter, Howard Baker to George Humphrey, June 13, 1956, PPRT, LC.

18. Letter, Sherman Adams to Charles Hook, June 22, 1956, PPRT, LC.

19. Letter, Lee Potter to Edward Burling, December 19, 1955, PPRT, LC; Letter, Russell Perkins to Leonard Hall, November 15, 1955, PPRT, LC; Letter, Russell Train to John Whitney, March 8, 1956, PPRT, LC.

20. Letter, David Kendall to Thomas Jenkins, June 8, 1956, PPRT, LC; Letter, Russell Train to Edward Burling, November 16, 1955, PPRT, LC; Letter, Russell Train to Gerald Morgan, July 3, 1956, PPRT, LC.

21. Letter, Russell Train to John P. Morgan, December 29, 1955, PPRT, LC; Letter, Leonard Hall to George Bowdoin, July 18, 1955, PPRT, LC; Letter, Charles Hook to Leonard Hall, June 19, 1956, PPRT, LC.

22. Letter, Russell Train to John Hollister, November 10, 1955, PPRT, LC.

23. Interview, Author with Russell Train, January 7, 2004; Letter, Russell Train to Charles Hook, August 28, 1956, PPRT, LC.

24. Interview, Author with Russell Train, January 7, 2004.

25. Russell Train, *A Memoir* (Washington, D.C.: self-published, 2000), 134.

26. Letter, Edward Murdock to Russell Train, April 24, 1957, PPRT, LC.

27. Interview, Author with Russell Train, January 7, 2004; Train, *A Memoir*, 139–141.

28. Letter, Russell Train to Grosvenor Chapman, January 18, 1969, PPRT, LC.

29. Interview, Author with Russell Train, January 7, 2004.

30. Ibid.

31. Letter, Rick Weyerhaeuser to Russell Train, July 19, 1994, PPRT, LC.

32. Interview, Author with Aileen Train, January 9, 2004.

33. Letter, Russell Train to Charles Hook, August 28, 1956, PPRT, LC.

34. Diary of Russell Train, 1956 Africa Trip, September 16 and 20, 1956, October 5 and 12, 1956, PPRT, LC.

35. Ibid., September 21 and 25, 1956.

36. Diary of Russell Train, 1958 Africa Trip, July 10, 1956, PPRT, LC.

37. Interview, Author with Russell Train, January 7, 2004.

38. Diary, 1956 Trip, October 2, 1956, PPRT, LC.

39. Ibid., October 8, 1956, PPRT, LC.

40. Ibid., September 21, 1956, PPRT, LC.

41. Russell Train, in Forward of Jacqueline Russell, *Thirty-Five Years of Conserving Wildlife in Africa: A History of the African Wildlife Foundation* (Washington, D.C.: African Wildlife Foundation, 1996).

42. Robin Hallet, *Africa since 1875* (Ann Arbor: University of Michigan Press, 1974), 571–587; Roland Oliver, *The African Experience* (New York, N.Y.: HarperCollins, 1991), 213–240.

43. Interview, Author with Russell Train, January 7, 2004.

44. Ibid.

45. Ibid.; Address by Russell Train, May 9, 1966, Ann Arbor, Michigan, PPRT, LC.

46. Typed Recollections of Hugh Lamprey, undated, PPRT, LC; Transcript, Interview, Harold K. Steen, Forest History Society, with Russell Train, 1993, PPRT, LC.

47. Interview, Author with Russell Train, January 7, 2004.

48. Ibid.

49. Ibid., January 8, 2004; Memorandum, Russell Train to Trustees, African Wildlife Leadership Fund, October 24, 1963, PPRT, LC.

50. Peter Scott, ed., *The Launching of the New Ark: The First Report of the President and Trustees of the World Wildlife Fund* (London, England: Collins, 1965), 15.

51. Ibid.; *New York Times*, June 8, 1962, 25.

52. Russell, *Thirty-Five Years*, 4–5.

53. Diary of Russell Train, 1963 East African Trip for the African Wildlife Leadership Fund, September 16, 1963, and October 5, 1963, PPRT, LC.

54. Letter, Rick Weyerhaeuser to Russell Train, September 20, 1994, PPRT, LC.

55. Interview, Author with Russell Train, January 7, 2004.

56. Ibid.

57. Letter, John Rhea to Russell Train, January 31, 1969, PPRT, LC.

58. Letter, Russell Train to Laurance Rockefeller, January 18, 1969, PPRT, LC; Letter, Russell Train to Lee Crandall, January 18, 1969, PPRT, LC; Biographical Summary, Russell Train, 1969, PPRT, LC.

59. Transcript, Lecture, Russell Train, "The Role of Foundations and Universities in Conservation," May 16, 1967, Berkeley, California, PPRT, LC.

60. Fairfield Osborn, *Our Plundered Planet* (Boston, Mass.: Little, Brown, 1948).

61. Quoted in Train, *Pollution, Pandas, and Politics*, 47; Interview, Author with Russell Train, January 7, 2004.

62. John Opie, *Nature's Nation: An Environmental History of the United States* (New York, N.Y.: Harcourt Brace, 1998), 422.

63. Transcript, Speech, Russell Train, "Open Space for Crowded Living," May 16, 1968, Kansas City, Missouri, PPRT, LC.

64. Train, *Pollution, Pandas, and Politics*, 48–49; Interview, Author with Russell Train, January 7, 2004.

65. Interview, Author with Aileen Train, January 9, 2004.

66. Interview, Author with Russell Train, January 7, 2004.

67. Quoted in Train, *A Memoir*, 174.

CHAPTER 3

1. John Kenneth Galbraith, *The Affluent Society* (Boston, Mass.: Houghton Mifflin, 1958).

2. Rachel Carson, *Silent Spring* (Boston, Mass.: Houghton Mifflin, 1962); for a view of her impact, see Linda Lear, *Rachel Carson: Witness for Nature* (New York, N.Y.: Holt, 1997).

3. Lewis L. Gould, *Lady Bird Johnson and the Environment* (Lawrence: University of Kansas Press, 1988), 54–55, 71–74, 121–123, 221–222.

4. Interview, Author with Russell Train, January 8, 2004.

5. Robin W. Winks, *Laurance S. Rockefeller: Catalyst for Conservation* (Washington, D.C.: Island Press, 1997), 70, 142–143.

6. Interview, Author with Russell Train, January 9, 2004.

7. Transcript, Address, Russell Train, Ann Arbor, Michigan, May 9, 1966, PPRT, LC.

8. Transcript, Address, Russell Train, Boston, Massachusetts, May 23, 1966, PPRT, LC.

9. John McPhee, *Encounters with the Archdruid* (New York, N.Y.: Farrar, Straus and Giroux, 1971), 87.

10. *Audubon* 86, no. 4 (September 1982): 469; Russell E. Train, *Politics, Pollution, and Pandas* (Washington, D.C.: Island Press, 2003), xiii.

11. Letter, Marion O'Connell to Russell Train, July 15, 1994, PPRT, LC.

12. Russell Train, *A Memoir* (Washington, D.C.: self-published, 2000), 174.

13. Transcript, Address, Russell Train, Kansas City, Missouri, May 16, 1968, PPRT, LC.

14. Interview, Author with Russell Train, January 8, 2004.

15. Train quoted in U.S. Congress, Senate Subcommittee on Merchant Marine and Fisheries, *Hearings on S.2984, A Bill to Prevent Importation of Endangered Species of Fish and Wildlife into the United States*, 90th Cong., 2nd sess., July 24, 1968, 56.

16. Train quoted in U.S. Congress, Senate Committee on Interior and Insular Affairs, *Hearings on S.1401, A Bill to Amend the Land and Water Conservation Fund Act*, 90th Cong., 2nd sess., February 8, 1968, 81.

17. *Science Teacher* 34, no.4 (April 1967): 22–24.

18. Conferences covered such topics as metropolitan development and "The Future of Environments in North America"; Ian L. McHarg, *Design with Nature* (Garden City, N.Y.: Natural History Press, 1969); Daniel Price, *The 99th Hour: The Population Crisis and the United States* (Chapel Hill: University of North Carolina Press, 1967); Anthony Netboy, *The Atlantic Salmon: A Vanishing Species?* (London, England: Faber, 1968).

19. Eddy's documentary also received funding from the Ford Foundation.

20. Train, *A Memoir*, 166.

21. Ibid., 231–234.

22. Ibid., 247–257.

23. Ibid., 241–246.

24. Diary of Russell Train, 1966 Trip to East Africa and Ethiopia, July 29 and August 30, PPRT, LC.

25. Journal of Aileen Train, 1966 African Trip, August 6–9, 1966, PPRT, LC.

26. Ibid., August 12, 1966, PPRT, LC.

27. "Recollections of Russell Train," Hugh Lamprey, undated, PPRT, LC.

28. Transcript, "The World Heritage Trust: Its Concept and Significance," Remarks of Russell Train, Washington, D.C., October 10, 1987, PPRT, LC.

29. Transcript, Address by Russell E. Train, International Congress on Man and Nature, Amsterdam, Netherlands, April 29, 1967, PPRT, LC.

30. Interview, Author with Russell Train, January 8, 2004.

31. J. Brooks Flippen, *Nixon and the Environment* (Albuquerque: University of New Mexico Press, 2000), 5.

32. James McEvoy, "The American Concern for the Environment," in William Burch et al., eds., *Social Behavior, Natural Resources, and the Environment* (New York, N.Y.: Harper and Row, 1972), 214–236; John Whitaker, *Striking a Balance: Environment and Natural Resources Policy in the Nixon-Ford Years* (Washington, D.C.: American Enterprise Institute, 1976), 7–9.

33. Interview, Author with Joe Browder, March 24, 1998.

34. Interview, Author with Russell Train, January 8, 2004.

35. *Living Wilderness* 32, no. 101 (Spring 1968): 2.

36. Interview, Author with Philip Berry, June 19, 1998.

37. Interview, Author with David Brower, April 7, 1998.

38. William C. Chafe, *The Unfinished Journey* (New York, N.Y.: Oxford University Press, 1986), 117, 123; Arthur S. Link and William B. Catton, *American Epoch* (New York, N.Y.: Alfred A. Knopf, 1980), 562, 566–567, 608–609.

39. Flippen, *Nixon and the Environment*, 2–3.

40. Joseph M. Petulla, *American Environmental History* (San Francisco, Calif.: Boyd and Fraser, 1977), 375; Victor B. Scheffer, *The Shaping of Environmentalism in America* (Seattle: University of Washington Press, 1998), 152.

41. The Air Quality Act of 1967 and the Water Quality Act of 1965.

42. Land and Water Conservation Act of 1965, the Solid Waste Disposal Act of 1965, and the Endangered Species Preservation Act of 1966; for an overview of the Johnson administration, see Doris Kearns, *Lyndon Johnson and the American Dream* (New York, N.Y.: Harper and Row, 1976).

43. *Sierra Club Bulletin* 53, no.1 (January 1968): 7; for more on Udall, see Thomas G. Smith, "John Kennedy, Stewart Udall, and the New Frontier Conservation," *Pacific Historical Review* 64, no. 3 (August 1995): 329.

44. Flippen, *Nixon and the Environment*, 7; for an overview of Muskie's career, see David Nevin, *Muskie of Maine* (New York, N.Y.: Random House, 1972), and Theo Lippman, Jr., and Donald C. Hansen, *Muskie* (New York, N.Y.: Norton, 1971).

45. Samuel Hays, *Beauty, Health, and Permanence: Environmental Politics in the United States, 1955–1985* (New York, N.Y.: Cambridge University Press, 1987), 43–52.

46. See Elmo Richardson, *Dams, Parks, and Politics: Development and Preservation in the Truman-Eisenhower Era* (Lexington: University of Kentucky Press, 1973).

47. Interview, Author with Russell Train, January 8, 2004.

48. United States National Water Commission, *Annual Report, 1969* (Washington, D.C.: U.S. Government Printing Press, 1969).

49. John Opie, *Nature's Nation: An Environmental History of the United States* (New York, N.Y.: Harcourt Brace, 1998), 324.

50. Joel Primack and Frank Von Hippel, "Scientists, Politics and the SST: A Critical Review," *Science and Public Affairs* 28, no. 4 (April 1972): 24–30; see also Melvin Horwitch, *Clipped Wings: The American SST Conflict* (Cambridge, Mass.: MIT Press, 1982).

51. Transcript, Interview, Harold K. Steen, Forest History Society, with Russell Train, 1993, 6–7, PPRT, LC.

52. Interview, Author with Russell Train, January 8, 2004.

53. Stephen Ambrose, *Nixon: The Triumph of a Politician, 1962–1972* (New York, N.Y.: Simon and Schuster, 1989), chapter 10.

54. Samuel Kirkpatrick and Melvin Jones, "Vote Direction and Issue Cleavage in 1968," *Social Science Quarterly* 51 (December 1970): 689–705; Philip Converse et al., "Continuity and Change in American Politics: Parties and Issues in the 1968 Election," *American Political Science Review* 63 (December 1969): 1083–1105; see David Caute, *The Year of the Barricades* (New York, N.Y.: Harper and Row, 1988), for a discussion of 1968.

55. Donald Johnson and Kirk Porter, *National Party Platforms* (Urbana: University of Illinois Press, 1973), 591–593, 612, 739, 758; Carl Solberg, *Hubert Humphrey* (New York, N.Y.: Norton, 1984), 419; Whitaker, *Striking a Balance*, 1–2.

56. Interview, Author with Russell Train, January 8, 2004.

57. Flippen, *Nixon and the Environment*, 17–18; see Roger Morris, *Richard Milhous Nixon: The Rise of an American Politician* (New York, N.Y.: Holt, 1990), for an overview of Nixon's background.

58. Interview, Author with Russell Train, January 8, 2004.

59. Membership List, Natural Resources and Environment Task Force, 1969, Folder "Task Force Reports, Transition Period, 1968–1969," Box 1, Transition Task Force Reports, White House Central Files (hereafter WHCF), RNPMP.

60. Interview, Author with Russell Train, January 8, 2004.

61. Report of the Natural Resources and Environment Transition Task Force, December 5, 1968, Folder "Task Force Reports, Transition Period, 1968–1969," Box 1, Transition Task Force Reports, WHCF, RNPMP.

62. Ibid., Appendix 2; Interview, Author with Russell Train, January 8, 2004.

63. J. Brooks Flippen, "Mr. Hickel Goes to Washington," *Alaska History* 12, no.2 (Fall 1998): 1–22; *New York Times*, December 20, 1968, 46.

64. Transcript, Interview, Harold K. Steen, Forest History Society, with Russell Train, 1993, 10–11, PPRT, LC.

65. *Congressional Record*, 91st Cong., 1st sess., vol. 115, pt. 2, January 21, 1969, 137.

66. Interview, Author with Russell Train, January 8, 2004.

67. Quoted in Train, *Politics, Pollution, and Pandas*, 8–9.

68. Interview, Author with Russell Train, January 8, 2004.

69. Ibid.; quoted in Train, *Politics, Pollution, and Pandas*, 10–11.

CHAPTER 4

1. *Public Papers of the Presidents: Richard Nixon, 1969* (Washington, D.C.: U.S. Government Printing Office, 1969–1974), 1–4.

2. Julie Nixon Eisenhower, *Pat Nixon: The Untold Story* (New York, N.Y.: Simon and Schuster, 1986), 252.

3. Walter Hickel, *Who Owns America?* (Englewood Cliffs, N.J.: Prentice Hall, 1971), 1–10.

4. U. S. Congress, Senate Committee on Interior and Insular Affairs, *Hearings on the Nomination of Russell E. Train to be Under Secretary of Interior*, 91st Cong., 1st sess., February 4, 1969, 3, 5.

5. Ibid., 6–7.

6. Letter, George Hartzog to Russell Train, July 4, 1994, PPRT, LC.

7. Transcript, Interview, Harold K. Steen, Forest History Society, with Russell Train, 1993, 11–12, PPRT, LC.

8. *New York Times*, February 2, 1969, 1; *Sierra Club Bulletin* 54, no. 3 (March 1969): 9.

9. Interview, Author with Russell Train, January 8, 2004.

10. Memo, Russell Train to Richard Nixon, March 27, 1969, PPRT, LC.

11. Memorandum for the Files, Russell Train, February 10, 1969, PPRT, LC.

12. *Santa Barbara News Release,* March 19, 1969, 8; *Los Angeles Times,* January 30, 1970, 1; Robert Easton, *Black Tide: The Santa Barbara Oil Spill and Its Consequences* (New York, N.Y.: Delacorte Press, 1972), 50–53, 134–136.

13. Quoted in Transcript, Hearings of Subcommittee on Air and Water Pollution, Senate Committee on Public Works, February 5, 1969, Folder 6, "1969 Correspondence, 500 Public Works, Santa Barbara Materials," Box 1560, United States Senate Staff Office Files (hereafter USSSO), Edmund Muskie Archives (hereafter EMA), Bates College, Lewiston, Maine.

14. Press Release, Senator Alan Cranston, February 26, 1969, Folder 2, "1969 Correspondence, 500 Public Works, Santa Barbara Materials," Box 1560, USSSO, EMA; Resolution, The Sierra Club, February 26, 1969, Folder 5, "1969 Correspondence, 500 Public Works, Santa Barbara Materials," Box 1560, USSSO, EMA.

15. Interview, Author with Walter Hickel, March 27, 1998; Memo, Walter Hickel to Richard Nixon, June 20, 1969, Folder "President's Handwriting, July, 1969," Box 2, President's Office Files, WHSF, RNPMP.

16. Interview, Author with Walter Hickel, March 27, 1998.

17. Interview, Author with Russell Train, January 8, 2004.

18. Transcript, Interview, Harold K. Steen, Forest History Society, with Russell Train, 1993, 13, PPRT, LC.

19. *Washington Daily News,* February 11, 1969, 24; *New York Times,* February 11, 1969, 31; *Washington Evening Star,* February 11, 1969, 12.

20. Memo, James Watt to Russell Train, February 24, 1969, PPRT, LC.

21. Executive Order 11472; Robert A. Shanley, "Presidential Executive Orders and Environmental Policy," *Presidential Studies Quarterly* 13, no.1 (Summer 1983): 405–416; Interview, Author with Russell Train, January 8, 2004.

22. Interview, Author with Russell Train, January 8, 2004; Jackson quoted in U.S. Congress, Senate Committee in Interior and Insular Affairs, *Hearings on S.1075 before the Senate Committee on Interior and Insular Affairs,* 91st Cong., 1st sess., June 1969, 137.

23. Nixon quoted in Memorandum for the President's File, Lee DuBridge, August 26, 1969, Folder "Meetings File, Beginning August 24, 1969," Box 79, President's Office Files, WHSF, RNPMP.

24. Interview, Author with John Whitaker, July 23, 1996; Interview, Author with Russell Train, January 8, 2004.

25. Memo, Russell Train to Assistant Secretaries, Solicitor, and Science Advisor, March 3, 1969, PPRT, LC.

26. Memo, Russell Train to Walter Hickel, June 26, 1969, PPRT, LC.

27. Memo, Russell Train to Walter Hickel, April, 1, 1969, PPRT, LC.

28. Memo, Russell Train to Assistant Secretary for Public Land Management et al., June 10, 1969, PPRT, LC.

29. Russell E. Train, *Politics, Pollution, and Pandas* (Washington, D.C.: Island Press, 2003), 61.

30. Memo, Richard Nixon to Walter Hickel, May 9, 1969, Folder "EX FG 221–25, Protection of Alaska's Arctic Environment, 1969–1970," Box 5, Presidential Task Forces, WHSF, RNPMP.

31. Letter, Russell Train to Thomas Kimball, May 9, 1969, PPRT, LC.

32. Letter, Russell Train to Burke Riley, May 19, 1969, PPRT, LC.

33. Letter, Russell Train to Stanley Powell, Jr., September 5, 1969, PPRT, LC.

34. Memo, Russell Train to Walter Hickel, July 23, 1969, PPRT, LC.

35. Peter Coates, *The Trans-Alaska Pipeline Controversy* (Bethlehem, Pa.: Lehigh University Press, 1991), 180–182; Alaska Pipeline Chronology of Events, 1968–1970, Folder "Alaska Pipeline, 1969–1970, 2 of 4," Box 25, John Whitaker Files, WHCF, RNPMP; Hickel, *Who Owns America?* 123–124.

36. Letter, Russell Train to Gardner Stout, March 3, 1969, PPRT, LC.

37. Memo, Russell Train to Walter Hickel, November 4, 1969, PPRT, LC.

38. Memo, Russell Train to Daniel Patrick Moynihan, May 9, 1969, PPRT, LC; Final Draft, Population Message to Congress, undated, Folder "White House—Environment, 1969," Box 9, Edward David Files, WHCF, RNPMP; *New York Times*, July 19, 1969, 1.

39. Memo, Russell Train to Carl McMurray, February 12, 1969, PPRT, LC.

40. Memo, Russell Train to Assistant Secretary, Public Land Management, July 22, 1969, PPRT, LC.

41. Memo, Russell Train to Assistant Secretary, Fish and Wildlife, Parks and Marine Resources, April 1, 1969, PPRT, LC; Memo, Russell Train to Assistant Secretary, Mineral Resources, June 16, 1969, PPRT, LC.

42. Memo, Russell Train to Assistant Secretary, Public Land Management, November 3, 1969, PPRT, LC.

43. Memo, Russell Train to Assistant Secretary, Water and Power Development, et al., June 26, 1969, PPRT, LC.

44. Flippen, *Nixon and the Environment*, 41; see also Ann Vileisis, *Discovering the Unknown Landscape: A History of America's Wetlands* (Washington, D.C.: Island Press, 1997).

45. Memo, Russell Train to Legislative Counsel, September 15, 1969, PPRT, LC.

46. Introduced as HR 14845; David Adams, "Management Systems under Consideration at the Federal-State Level," in James C. Hite and James M. Stepp, *Coastal Zone Resource Management* (New York, N.Y.: Praeger, 1971), 23–29.

47. P.L. 92–583; Robert B. Abel, "The History of the United States Ocean Policy Program," in Francis Hoole, Robert Friedheim, and Timothy Hennessey, eds., *Making Ocean Policy* (Boulder, Colo.: Westview Press, 1981), 27; John Whitaker, *Striking a Balance: Environment and Natural Resources Policy in the Nixon-Ford Years* (Washington, D.C.: American Enterprise Institute, 1976), 153–154.

48. Interview, Author with Maurice Stans, August 11, 1996; Memo, Russell Train to Walter Hickel, September 30, 1969, PPRT; Memo, Russell Train to Walter Hickel, November 3, 1969, PPRT, LC; Interview, Author with Russell Train, January 8, 2004.

49. Russell Train, *A Memoir* (Washington, D.C.: self-published, 2000), 184; Memo, Russell Train to Walter Hickel, October 3, 1969, PPRT, LC; Memo, Russell Train to Director, Bureau of Commercial Fisheries, et al., October 30, 1969, PPRT, LC.

50. Letter, Russell Train to Alan Bible, May 9, 1969, PPRT, LC.

51. Memo, Russell Train to Assistant to the Secretary and Director of Information, August 8, 1969, PPRT, LC; Memo, Russell Train to Walter Hickel, October 28, 1969, PPRT, LC; Memo, Russell Train to Assistant Secretary, Water and Power Development, et al., August 8, 1969, PPRT, LC.

52. Nixon quoted in Diary, Russell Train, 1969–1970 Journal, February 7, 1970, PPRT, LC; Agenda, End of the Year Briefings, December 19, 1969, PPRT, LC.

53. Letter, Edward Gurney to Russell Train, October 30, 1969, PPRT, LC; Letter, Ted Stevens to Russell Train, November 13, 1969, PPRT, LC; Letter, Russell Train to Karl Mundt, July 14, 1969, PPRT, LC.

54. Letter, Russell Train to Sydney Howe, December 1, 1969, PPRT, LC; Letter, Russell Train to Charles Lindbergh, October 27, 1969, PPRT, LC; Memo, Russell Train to Walter Hickel, Responses to Questions Posed for *National Wildlife* Article, undated, 1969, PPRT, LC.

55. Train, *Politics, Pollution, and Pandas*, 65–67; Interview, Author with Russell Train, January 8, 2004.

56. Letter, Russell Train to Takeso Shimoda, November 13, 1969, PPRT, LC; Memo, Russell Train to Walter Hickel, August 12, 1969, PPRT, LC.

57. *Public Papers of the Presidents: Richard Nixon, 1969*, 223; Russell Train, "A New Approach to International Environmental Cooperation: The NATO Committee on the Challenges of Modern Society," *Kansas Law Review* 22, no. 21 (Winter 1974): 167–191.

58. Memo, Russell Train to Daniel P. Moynihan, June 26, 1969, PPRT, LC; Letter, Walter Hickel to Armin H. Meyer, August 8, 1969, PPRT, LC.

59. Quoted in *Seattle Post-Intelligencer*, July 20, 1969, A1; Howard Bloomfield, "The Everglades: Pregnant with Risks," *American Forests* 78 (May 1970): 27; Gary Soucie, "The Everglades Jetport: One Hell of an Uproar," *Sierra Club Bulletin* 54 (July 1969): 23; *BioScience* 19, no. 7 (July 1969): 64; Walter Rosenbaum, *The Politics of Environmental Concern* (New York, N.Y.: Praeger, 1974), 168–196.

60. Transcript, Interview, Harold K. Steen, Forest History Society, with Russell Train, 1993, 14, PPRT, LC.

61. Memo, Russell Train to Luna Leopold, October 1, 1969, PPRT, LC; Memo, Russell Train to Walter Hickel, December 9, 1969, PPRT, LC.

62. Interview, Author with Russell Train, January 8, 2004.

63. Ibid.; *Wall Street Journal*, June 18, 1969, 1.

64. Diary of Russell Train, 1969–1970 Journal, September 4 and September 17, 1969, PPRT, LC.

65. Richard N. L. Andrews, *Environmental Policy and Administrative Change* (Lexington, Mass.: D.C. Heath, 1976), 9–12; Interview, Author with Leon Billings, June 26, 1998; Muskie termed his bill the Water Quality Improvement Act.

66. U.S. Congress, Senate Committee on Interior and Insular Affairs, *Hearings before the Senate Committee on Interior and Insular Affairs on S.1075, S.237and S.1725*, 91st Cong., 1st sess., April 16, 1969, 73–76; Transcript, Interview, Harold K. Steen, Forest History Society, with Russell Train, 1993, 9, PPRT, LC.

67. Memo, Russell Train to John Ehrlichman, November 3, 1969, Folder "485–OA 2977, Environment," Box 63, Egil Krogh Files, WHSF, RNPMP.

68. Memo, Russell Train to Lee DuBridge, May 9, 1969, PPRT, LC.

69. Memo, Russell Train to Science Advisor, July 22, 1969, PPRT, LC.

70. Interview, Author with Russell Train, January 8, 2004.

71. Memo, Russell Train to John Whitaker, January 8, 1969, PPRT, LC.

72. Diary of Russell Train, 1969–1970 Journal, February 7, 1970, PPRT, LC.

73. P.L. 91–190; *Public Papers of the Presidents: Richard Nixon, 1970*, 2–3; *New York Times*, January 2, 1970, 1.

74. Diary of Russell Train, 1969–1970 Journal, February 7, 1970, PPRT, LC; Interview, Author with Russell Train, January 8, 2004.

CHAPTER 5

1. Quoted in Ashton Applewhite et al., *And I Quote* (New York, N.Y.: St. Martin's Press, 2003), 18.

2. *Washington Post,* January 30, 1970, 1; *New York Times,* January 30, 1970, 1; *Time* (February 9, 1970): 46; *Business Week* (April 18, 1970): 136; Diary of Russell Train, 1969–1970 Journal, February 7, 1970, PPRT, LC.

3. *Washington Post,* January 30, 1970, 1; *Philadelphia Inquirer,* February 4, 1970; *Sunday Herald Traveler,* March 8, 1970, 51; *Charlotte Observer,* January 30, 1970, 92; *Spokesman Review,* February 2, 1970, 41.

4. *Arizona Republic,* January 16, 1970, 21.

5. Press Release, Environmental Message, February 10, 1970; Folder "Environmental Message, February 10, 1970," Box 58, John Whitaker Files, WHCF, RNPMP; *Public Papers of the Presidents: Richard Nixon, 1970* (Washington, D.C.: U.S. Government Printing Office, 1969–1974), 95–108; *New York Times,* February 11, 1970, 1.

6. Russell E. Train, *Politics, Pollution, and Pandas* (Washington, D.C.: Island Press, 2003), 83.

7. U.S. Congress, Senate Committee on Interior and Insular Affairs, *Hearings before the Senate Committee on Interior and Insular Affairs on the Nomination of Russell E. Train to the Council on Environmental Quality,* 91st Cong., 2nd sess., February 5, 1970; quoted in Diary of Russell Train, 1969–1970 Journal, February 7, 1970, PPRT, LC.

8. Memo, Russell Train to Richard Nixon, August 12, 1970, Folder "Supersonic Transport, 1969–1970, 2 of 6," Box 109, John Whitaker Files, WHCF, RNPMP; Joel Primack and Frank Von Hippel, "Scientists, Politics and the SST: A Critical Review," *Science and Public Affairs* 28, no. 4 (April 1972): 28; Interview, Author with Russell Train, January 8, 2004.

9. Interview, Author with Russell Train, January 8, 2004; Train, *Politics, Pollution, and Pandas,* 85.

10. Memo, Russell Train to John Ehrlichman, July 23, 1970, PPRT, LC; Memo, Russell Train to John Whitaker, October 10, 1970, PPRT, LC.

11. *Washington Star,* January 11, 1970, A2; Memo, Russell Train to Robert Cahn and Gordon MacDonald, March 16, 1970, PPRT, LC; Memorandum for the Files, Russell Train, February 13, 1970, PPRT, LC; Letter, Russell Train to Laurance Rockefeller, November 19, 1970, PPRT, LC.

12. Memorandum for the Files, Russell Train, February 18, 1970, PPRT, LC; Memorandum for the Files, Russell Train, March 19, 1970, PPRT, LC; Memorandum for the Files, Robert Cahn, February 21, 1970, PPRT, LC.

13. Letter, Russell Train to John Mitchell, April, 2, 1970, PPRT, LC; Memo, Russell Train to Tim Atkeson, March 9, 1970, PPRT, LC; Memo, Russell Train to John Whitaker, April 13, 1970, PPRT, LC; Letter, Russell Train to Frank Craighead, March 17, 1970, PPRT, LC.

14. Letter, Gordon MacDonald to Kathryn Fuller, July 6, 1994, PPRT, LC.

15. Memo, Russell Train to Robert Mayo, April 27, 1970, PPRT, LC.

16. Letter, Frank Hodsell to Russell Train, July 18, 1994, PPRT, LC.

17. Memo, Russell Train to John Ehrlichman, April 4, 1970, PPRT, LC; Memo, John Ehrlichman to Harry Fleming, undated, PPRT, LC.

18. Letter, Terry Davies to Kathryn Fuller, July 13, 1994, PPRT, LC.

19. Memo, Russell Train to John Ehrlichman, March 23, 1970, PPRT, LC; Marc Landy, Marc Roberts, and Stephen Thomas, *The Environmental Protection Agency: Asking the Wrong Questions; From Nixon to*

Clinton (New York, N.Y.: Oxford University Press, 1994), 31–32; Chronology of Ash Council—Natural Resources, December 1971, Folder "Chronology of Reorganization Proposals," Box 22, PACEO Files, WHCF, RNPMP.

20. Memo, Russell Train to John Ehrlichman, April 27, 1970, PPRT, LC.

21. Interview, Author with Russell Train, January 8, 2004.

22. Memo, Russell Train to Gordon MacDonald, April 28, 1970, PPRT, LC; Memo, Lee DuBridge to Director, Bureau of the Budget, et al., May 2, 1970, PPRT, LC.

23. Memo, Al Alm to Staff Group on National Environmental Institute, May 27, 1970, PPRT, LC; Memo, Russell Train to George Schultz, May 13, 1971, PPRT, LC; Council on Environmental Quality, *The President's 1971 Environmental Program* (Washington, D.C.: U.S. Government Printing Office, 1971), 87.

24. Memo, Maurice Stans to Richard Nixon, February 4, 1970, Folder "White House—Issues, 1970, 1 of 2," Box 11, Edward David Files, WHCF, RNPMP; Memo, Maurice Stans to Richard Nixon, February 4, 1970, Folder "NIPCC, 1970–71, 5 of 5," Box 84, John Whitaker Files, WHCF, RNPMP.

25. Memorandum for the President's Files, John Whitaker, February 4, 1970, Folder "Meetings File, Beginning February 1, 1970," Box 80, President's Office Files, White House Special Files (hereafter WHSF), RNPMP.

26. Memo, John Whitaker to Maurice Stans, February 5, 1970, Folder "NIPCC, 1970–71, 5 of 5," Box 84, John Whitaker Files, WHCF, RNPMP.

27. Memo, Russell Train to Hugh Sloan, October 23, PPRT, LC.

28. Executive Order 11514; Robert A. Shanley, *Presidential Influence and Environmental Policy* (Westport, Conn.: Greenwood Press, 1991), 53; Press Release, Executive Order, Protection and Enhancement of Environmental Quality, March 5, 1970, Folder "The Environmental Coalition, 1 of 2," Box 61, Charles Colson Files, WHSF, RNPMP; see also *Government Executive* (June 1970): 65–67.

29. Letter, Tim Atkeson to Kathryn Fuller, undated, PPRT, LC.

30. *Life* (August 28, 1970): 31–34; Interview, Author with Russell Train, January 8, 2004.

31. Quoted in Train, *Politics, Pollution, and Pandas*, 21; Memo, Russell Train to Heads of Executive Departments and Establishments, April 29, 1970, PPRT, LC.

32. Memo, Russell Train to Tim Atkeson, May 19, 1970, PPRT, LC.

33. Memo, Russell Train to Boyd Gibbons, April 6, 1970, PPRT, LC.

34. Memo, Russell Train to Henry Jackson, November 9, 1970, PPRT, LC; Memo, Russell Train to John Whitaker, November 24, 1970, PPRT, LC; Letter, Russell Train to M. B. Lane, November 24, 1970, PPRT, LC.

35. Transcript, Interview with Russell Train, WNBC-Television, Speaking Freely, August 17, 1970, PPRT, LC; Train, *Politics, Pollution, and Pandas*, 90–91.

36. Train, *Politics, Pollution, and Pandas*, 89; *Chicago Tribune*, February 5, 1970, 1.

37. *New York Times*, April 23, 1970, 1; *Washington Post*, April 23, 1970, 1; Steve Cotton, "What Happened," *Audubon* 72, no. 4 (July 1970): 112–115.

38. Memo, John Whitaker to Christopher DeMuth, February 2, 1970, Folder "January–April, 1970, 2 of 4, February, 1970," Box 2, John Whitaker Files, WHCF, RNPMP; J. Brooks Flippen, *Nixon and the Environment* (Albuquerque: University of New Mexico Press, 2000), 8–13.

39. Letter, Boyd Gibbons to Charles Hitch, February 20, 1970, PPRT, LC; Letter, Russell Train to Robert Anderson, March 10, 1970, PPRT, LC; Train quoted in *Orlando Sentinel*, May 5, 1970, 1.

40. Interview, Author with Russell Train, January 8, 2004.

41. Nixon quoted in Haldeman Personal Notes, May 10, 1970, Folder "Haldeman Notes, April–June, 1970, Part II," Box 41, H. R. Haldeman Files, WHSF, RNPMP.

42. Hickel quoted in *Business Week* (February 14, 1970): 57.

43. Letter, Russell Train to Walter Hickel, January 30, 1970, PPRT, LC; Letter, Walter Hickel to Russell Train, May 22, 1970, PPRT, LC.

44. Memo, Russell Train to John Whitaker, June 4, 1970, PPRT, LC.

45. Memo, Russell Train to John Whitaker, July 16, 1970, PPRT, LC.

46. Council on Environmental Quality, *First Annual Report of the President's Council on Environmental Quality* (Washington, D.C.: U.S. Government Printing Office, 1970).

47. *New York Daily News*, August 17, 1970, 3; *New York Times*, August 11, 1970, 1.

48. Letter, Russell Train to John Lear, September 22, 1970, PPRT, LC.

49. Memo, Russell Train to Richard Nixon, December 1, 1970, PPRT, LC; Council on Environmental Quality, *Second Annual Report of the President's Council on Environmental Quality* (Washington, D.C.: U.S. Government Printing Office, 1971); Council on Environmental Quality, *The President's 1971 Environmental Program* (Washington, D.C.: U.S. Government Printing Office, 1971); *New York Times*, August 11, 1971, 36.

50. Quoted in Train, *Politics, Pollution, and Pandas*, 105.

51. Memo, Russell Train to John Whitaker, December 1, 1970, PPRT, LC; Memo, Russell Train to John Whitaker, April 14, 1970, PPRT, LC; Tim Palmer, *Endangered Rivers and the Conservation Movement* (Berkeley: University of California Press, 1986), 97; Victor B. Scheffer, *The Shaping of Environmentalism in America* (Seattle: University of Washington Press, 1998), 136.

52. Memo, Russell Train to John Whitaker, May 21, 1970, PPRT, LC; Letter, Russell Train to Claude Kirk, July 28, 1970, PPRT, LC; Letter, Russell Train to Marjorie Carr, April 21, 1970, PPRT, LC; Memo, Russell Train to John Whitaker, December 1, 1970, PPRT, LC; Press Release, White House Secretary, January 19, 1971, Folder "Cross Florida Barge Canal, 1970–1971, 2 of 4," Box 44, John Whitaker Files, WHCF, RNPMP.

53. Memo, Russell Train to John Whitaker, October 26, 1970, PPRT, LC; Memo, Russell Train to Richard Nixon, April 14, 1970, PPRT, LC; Janet Scheffer, "Solid Wastes," in James Rathlesberger, ed., *Nixon and the Environment* (New York, N.Y.: Village Voice, 1972), 239–241; Scheffer, *Shaping of Environmentalism*, 152; CEQ Decision Document, Automobile Cycle: Environmental and Resource Conservation Problems, June 1970, Folder "White House—Environment, Oversized, 1969–1970," Box 9, Edward David Files, WHCF, RNPMP; Memo, Russell Train to Richard Nixon, June 4, 1970, Folder "White House—Environment, Vol. III, 1971," Box 9, Edward David Files, WHCF, RNPMP.

54. Memo, Russell Train to Richard Nixon, April 30, 1970, PPRT, LC; Memo, Russell Train to Secretary of Treasury et al., May 5, 1970, PPRT, LC; Memo, Russell Train to Al Alm, May 25, 1970, PPRT, LC; Letter, Russell Train to Walter Washington, August 14, 1970, PPRT, LC; P.L. 91–512.

55. Press Release, Identification of Unneeded Federal Real Property, February 10, 1970, Folder "Parklands," Box 56, Edwin Harper Files, WHCF, RNPMP; Train, *Politics, Pollution, and Pandas*, 88.

56. *New York Times*, September 27, 1970, 1; September 28, 1970, 44; Press Release, Air Pollution, Backing Words with Action, Edmund Muskie, August 16, 1970, Folder 3, "1970 Press File Ledger Syndicate Columns, 1970—So Goes the Nation," Box 1533, USSSO, EMA; Interview, Author with Don Nicholl, April 1, 1998; Memo, John Whitaker to Ken Cole, September 1, 1970, Folder "September–December, 1970, 1 of 4, September, 1970," Box 2, John Whitaker Files, WHCF, RNPMP.

57. Memo, Russell Train to John Ehrlichman, December 14, 1970, PPRT, LC; Memo, Russell Train to Al Alm, August 17, 1970, PPRT, LC; Transcript, Testimony of the Honorable Russell E. Train before the Committee on Ways and Means, House of Representatives, in Support of the Administration's Request for Lead Additives Tax, September 9, 1970, PPRT, LC.

58. Memorandum for the Files, Russell Train, February 17, 1970, PPRT, LC; Memo, Russell Train to John Volpe, May 13, 1970, PPRT, LC.

59. Memo, Russell Train to Richard Nixon, October 1, 1970, PPRT, LC.

60. P.L. 91–604; Interview, Author with Leon Billings, June 26, 1998; Press Release, Muskie Commends Nixon for Signing Air Bill, December 31, 1970, Folder 14, "1970 Press File, December 31, 1970, Muskie Commends President, Air Bill," Box SE1537, USSSO, EMA; Interview, Author with Don Nicholl, April 1, 1998; *New York Times,* January 1, 1971, 1; *Washington Post,* January 1, 1971, 1; Memo, Charles Colson to Dwight Chapin, January 12, 1971, Folder "Air Pollution—General, 1 of 2," Box 30, Charles Colson Files, WHSF, RNPMP; Memo, Jeb Magruder to Ed Morgan, February 25, 1971, Folder "February, 1971, 1 of 2," Box 271, H. R. Haldeman Files, WHSF, RNPMP.

61. Congressional Research Service, *A Legislative History of the Water Pollution Control Act Amendments of 1972* (Washington, D.C.: U.S. Government Printing Office, 1973), 281–339; Harvey Lieber, *Federalism and Clean Waters: The 1972 Water Pollution Control Act* (Lexington, Mass.: D.C. Heath and Co., 1975), 12, 80–82; Water Quality Improvement Act of 1970, P.L. 91–224.

62. 30 Stat. 1151; Memo, Russell Train to John Mitchell et al., October 28, 1970, PPRT, LC; Memo, Russell Train to John Mitchell et al., September 14, 1970, PPRT, LC.

63. Memo, John Whitaker to John Ehrlichman, October 6, 1970, Folder "September–December, 1970, 2 of 4, October, 1970," Box 2, John Whitaker Files, WHCF, RNPMP; Interview, Author with Russell Train, January 8, 2004; *Miami Herald,* January 30, 1970, 1; *Washington Post,* December 24, 1970, 1.

64. Memo, Russell Train to John Dingell, September 29, 1970, PPRT, LC.

65. Council on Environmental Quality, *Ocean Dumping: A National Policy* (Washington, D.C.: U.S. Government Printing Office, 1970); Memo, Russell Train to Secretary of Interior et al., April 30, 1970, PPRT, LC; Train quoted in *Washington Post,* October 8, 1970, A8.

66. David Zwick, "Water Pollution," in Rathlesberger, *Nixon and the Environment,* 45–46; see Terence Kehoe, *Cleaning Up the Great Lakes: From Cooperation to Confrontation* (DeKalb: Northern Illinois University Press, 1997).

67. Train designated Gordon MacDonald as the official chair of the Joint Working Group; Memo, Russell Train to Richard Nixon, May 22, 1970, PPRT, LC; Memo, Russell Train to Al Alm, April 20, 1970, PPRT, LC; Memo, Russell Train to Secretary of Commerce et al., April 28, 1970, PPRT, LC.

68. Memo, Russell Train to William Rogers, December 24, 1970, PPRT, LC; Memo, Russell Train to Secretary of State et al., August 14, 1970, PPRT, LC; Memo, Russell Train to Richard Nixon, July 2, 1970, PPRT, LC.

69. Report of the U.S.-Canada Joint Working Group, June 10, 1971, Folder "Great Lakes, 1971–1972, 2 of 2," Box 70, John Whitaker Files, WHCF, RNPMP; *Washington Post,* June 11, 1971, 1.

70. See William Bundy, *A Tangled Web: The Making of Foreign Policy in the Nixon Presidency* (New York, N.Y.: Hill and Wang, 1998).

71. See Henry Kissinger, *White House Years* (Boston, Mass.: Little, Brown, 1979), and Richard C. Thornton, *The Nixon-Kissinger Years* (St. Paul, Minn.: Paragon, 2001).

72. Interview, Author with Russell Train, January 8, 2004.

73. Ibid.; Memo, Russell Train to William P. Rogers, April 8, 2004, PPRT, LC.

74. Interview, Author with Russell Train, January 8, 2004.

75. Letter, Russell Train to Nobuhiko Ushiba, October 8, 1970, PPRT, LC.

76. *Washington Post,* October 18, 1970, A4; *Mainichi Shimbun,* October 15, 1970, 1; *Asahi Evening News,* October 14, 1970, 3.

77. Interview, Author with Aileen Train, January 9, 2004; Train, *Politics, Pollution, and Pandas,* 148–149; Letter, Russell Train to Clinton Atkinson, PPRT, LC.

78. Interview, Author with Russell Train, January 8, 2004.

79. Train, *Politics, Pollution, and Pandas,* 149–150.

80. Letter, Russell Train to Jacques Duhamel, September 24, 1970, PPRT, LC; Memorandum for the Files, Boyd Gibbons, March 23, 1970, PPRT, LC.

81. Press Release, Russell Train to Head NATO CCMS, January 13, 1971, PPRT, LC; Letter, Richard Nixon to Russell Train, January 14, 1971, PPRT, LC; Interview, Author with Aileen Train, January 9, 2004.

82. Letter, Russell Train to Jeremiah Milbank, Jr., September 23, 1970, PPRT, LC; Memo, Russell Train to Ed Coate, November 10, 1970, PPRT, LC; Letter, Russell Train to Hugh Scott, November 5, 1970, PPRT, LC.

83. *Washington Post,* October 17, 1970, 1; October 14, 1970, 31; *New York Times,* October 10, 1970, 1; June 10, 1970, C30; Letter, Russell Train to Dan Smith, April 1, 1970, PPRT, LC; *Princeton Alumni Weekly* (June 30, 1970): 10; *American Forests* (June 1970): 5.

CHAPTER 6

1. Interview, Author with Russell Train, January 7, 2004.

2. Michael Schaller and George Rising, *The Republican Ascendancy: American Politics, 1968–2001* (Wheeling, Ill.: Harlan Davidson, 2002), 31–33.

3. Nixon quoted in Herbert Parmet, *Richard Nixon and His America* (Boston, Mass.: Little, Brown, 1990), 529; see also Richard Nixon, *RN: The Memoirs of Richard Nixon* (New York, N.Y.: Grosset and Dunlap, 1978).

4. Stephen Ambrose, *Nixon: The Triumph of a Politician, 1962–1972* (New York, N.Y.: Simon and Schuster, 1989), 238, 327.

5. Safire quoted in Schaller and Rising, *Republican Ascendancy,* 29.

6. E. J. Dionne, Jr., "In Nixon's Shadow," *Washington Post,* August 20, 1999, 35; for a solid overview of Nixon's career, see Jonathan Aitken, *Nixon: A Life* (Washington, D.C.: Regnery Publishing, 1993); for the best study of Nixon's political rise, see Roger Morris, *Richard Milhous Nixon: The Rise of an American Politician* (New York, N.Y.: Holt, 1990).

7. Broder quoted in Ambrose, *Nixon: The Triumph of a Politician,* 495.

8. Richard M. Scammon and Ben J. Wattenberg, *The Real Majority* (New York, N.Y.: Coward-Mc-Cann, 1970).

9. See Robert Mason, *Richard Nixon and the Quest for a New Majority* (Chapel Hill: University of North Carolina Press, 2004).

10. For a full discussion of Nixon's shifting attitudes on the environment, see J. Brooks Flippen, *Nixon and the Environment* (Albuquerque: University of New Mexico Press, 2000); Nixon quoted in Haldeman Personal Notes, February 9, 1971, Folder "Notes, January–March, 1971 (January 1–February 15, 1971), Part I," Box 43, H. R. Haldeman Files, WHSF, RNPMP.

11. Nixon quoted in Memorandum for the President's File, Charles Colson, March 9, 1971, Folder "Meetings File, Beginning March 7, 1971," Box 84, President's Office Files, WHSF, RNPMP; Interview, Author with Richard Fairbanks, March 24, 1998.

12. Letter, Maurice Stans to Richard Nixon, April 12, 1971, Folder "Meetings File, Beginning April 12, 1971," Box 84, President's Office Files, WHSF, RNPMP; Nixon quoted in Haldeman Personal Notes, June 9, 1971, Folder "Notes, April–June, 1971 (May 20–June 30, 1971), Part II," Box 43, H. R. Haldeman Files, WHSF, RNPMP.

13. Memo, Will Kriegsman to John Whitaker, March 29, 1971, Folder "January–March, 1971, 3 of 3, March, 1971," Box 3, John Whitaker Files, WHCF, RNPMP.

14. Memorandum for the Files, Russell Train, February 26, 1973, PPRT, LC.

15. *New York Times*, February 9, 1972, 1.

16. David Zwick, "Water Pollution," in James Rathlesberger, ed., *Nixon and the Environment* (New York, N.Y.: Village Voice, 1972), 43; Harvey Lieber, *Federalism and Clean Waters: The 1972 Water Pollution Control Act* (Lexington, Mass.: D.C. Heath and Co., 1975), 24–25.

17. Quoted in *Washington Post*, March 23, 1972, 34.

18. *Kalur v. Resor*, Fed. Supp. 1, 3 ERC1458, 1 ELR 20637 (D.D.C., 1971).

19. *Washington Evening Star*, February 2, 1971, 2.

20. Interview, Author with Russell Train, January 8, 2004.

21. Quoted in *New Republic* (January 22, 1972): 6–7; *Washington Post*, December 8, 1971, 6.

22. Quoted in *New York Times*, January 1, 1972, 1; *Water in the News* (September 1971): 1–2.

23. *New York Times*, October 19, 1972, 1, 46; Stanley Kutler, *The Wars of Watergate* (New York, N.Y.: Alfred A. Knopf, 1990), 135–136; John Ehrlichman, *Witness to Power: The Nixon Years* (New York, N.Y.: Pocket Books, 1982), 330; Lieber, *Federalism and Clean Waters*, 82.

24. Memorandum for the Files, Russell Train, February 3, 1972, PPRT, LC.

25. Samuel Hays, *Beauty, Health, and Permanence: Environmental Politics in the United States, 1955–1985* (New York, N.Y.: Cambridge University Press, 1987), 140; *New York Times*, September 12, 1972, 31.

26. Memorandum for the Files, Russell Train, June 21, 1973, PPRT, LC; Memorandum for the Files, Russell Train, June 22, 1973, PPRT, LC; Russell E. Train, *Politics, Pollution, and Pandas* (Washington, D.C.: Island Press, 2003), 94.

27. Memo, John Whitaker to Jeanne Davis, March 13, 1972, PPRT, LC; Memo, Henry Kissinger to Secretary of State, April 4, 1972, PPRT, LC; *Greencastle (Ind.) Banner Graphic*, September 25, 1971, 1; Terence Kehoe, *Cleaning Up the Great Lakes: From Cooperation to Confrontation* (DeKalb: Northern Illinois University Press, 1997), 149, 164–165; Flippen, *Nixon and the Environment*, 165; Train quoted in *U.S. News and World Report* (September 27, 1976): 51.

28. John Whitaker, *Striking a Balance: Environment and Natural Resources Policy in the Nixon-Ford Years* (Washington, D.C.: American Enterprise Institute, 1976), 147–150.

29. Memo, John Whitaker to John Ehrlichman, April 10, 1970, Folder "January–April, 1970, 4 of 4, April, 1970," Box 2, John Whitaker Files, WHCF, RNPMP.

30. Letter, Russell Train to Alan Cranston, May 15, 1970, PPRT, LC.

31. Letter, Russell Train to James Beggs, September 25, 1970, PPRT, LC.

32. Quoted in *Birmingham News*, June 19, 1973, 28.

33. Quoted in *Minneapolis Tribune*, March 3, 1972, B2.

34. Memo, John Whitaker to John Ehrlichman, June 29, 1970, Folder "May–August, 1970, 2 of 4, June, 1970," Box 2, John Whitaker Files, WHCF, RNPMP.

35. CEQ Draft Bill, National Land Use Policy, October 29, 1970, Folder "Land Use Policy, 1970," Box 15, Egil Krogh Files, WHSF, RNPMP; Whitaker, *Striking a Balance*, 156–157; Letter, Russell Train to George Schultz, December 18, 1970, PPRT, LC; *Washington Post*, December 8, 1971, 21.

36. Fred Bosselman and David Callies, *The Quiet Revolution in Land Use Control* (Washington, D.C.: U.S. Government Printing Office, 1971).

37. James Noone, "Senate, House Differ in Approaches to Reform of Nation's Land Use Laws," *National Journal* (July 22, 1972): 1193; Letter, Richard Nixon to Chairman Henry Jackson, April 24, 1972, Folder "EX NR Natural Resources, 2 of 3, 1971–1972," Box 1, Natural Resources Files, WHCF, RNPMP.

38. Quoted in *New York Times*, January 23, 1970, 1.

39. Whitaker, *Striking a Balance*, 163.

40. Ibid., 162–165; *Congressional Quarterly Almanac*, 93rd Cong., 2nd sess., vol. 30 (1974), 790; *Congressional Record*, 93rd Cong., 2nd sess., vol. 120, pt. 10 (June 11, 1974), H5028.

41. Memo, Ken Cole to Russell Train, October 9, 1973, Folder "EX FG 298 EPA," Box 2, EPA Files, WHCF, RNPMP.

42. Memo, Russell Train to Laurence Woodworth, December 16, 1970, PPRT, LC; *Washington Post*, June 19, 1973, 26; Memo, Russell Train to John Whitaker, March 10, 1970, PPRT, LC.

43. Memo, Russell Train to Secretary of Agriculture et al., July 15, 1970, PPRT, LC; Memorandum for the Files, Russell Train, February 17, 1970, PPRT, LC; *Des Moines Register*, February 12, 1971, 1; Poage quoted in Train, *Politics, Pollution, and Pandas*, 94.

44. Executive Order 11643, February 8, 1972; Thomas Dunlap, *Saving America's Wildlife* (Princeton, N.J.: Princeton University Press, 1988), 157; Press Release, Control of Predatory Animals, July 1974, Folder "GEN NR Fish—Wildlife, 17 of 17, June, 1973–July, 1974," Box 6, Natural Resources Files, WHCF, RNPMP.

45. Memo, Russell Train to Donald Rice, February 2, 1972, Folder "Off-Road Vehicles, 1972," Box 88, John Whitaker Files, WHCF, RNPMP; Letter, Russell Train to Guy Vander Jagt, April 4, 1970, PPRT, LC.

46. Executive Order 11644, February 8, 1972; Paul J. Culhane, *Public Lands Politics* (Baltimore, Md.: Johns Hopkins University Press, 1981), 171, 174.

47. See John F. Stacks, *Stripping: The Surface Mining of America* (San Francisco, Calif.: Sierra Club Books, 1972); Hays, *Beauty, Health, and Permanence*, 144–145; Train quoted in *Environmental Quality Magazine* (June 1973): 24.

48. *Sierra Club v. Hardin* 325 F. Supp. 99 (D. Alaska, 1971); David Clary, *Timber and the Forest Service* (Lawrence: University of Kansas Press, 1986), 180–181; *New York Times*, January 18, 1972, 24.

49. See J. Brooks Flippen, "The Nixon Administration, Timber and the Call of the Wild," *Environmental History Review* 19, no. 2 (Summer 1995): 37–54; Train quoted in Memo, John Whitaker to Russell Train, December 4, 1969, Folder "June–December, 1969, 6 of 6, December, 1969," Box 1, John Whitaker Files, WHCF, RNPMP.

50. Public Land Law Review Commission, *One Third of the Nation's Land: A Report to the President and Congress* (Washington, D.C.: U.S. Government Printing Office, 1970); Letter, Russell Train to Jean Thomas, September 24, 1970, PPRT, LC; Memo, Russell Train to Boyd Gibbons, March 9, 1970, PPRT, LC; Carl Mayer and George A. Riley, *Public Domain, Private Dominion* (San Francisco, Calif.: Sierra Club Books, 1985), 11–12.

51. Interview, Author with Russell Train, January 8, 2004.

52. Letter, Russell Train to Elliot Richardson, November 17, 1969, PPRT, LC; Memo, Russell Train to Henry Kissinger, December 3, 1970, PPRT, LC.

53. Transcript, Statement of Russell Train before the Subcommittee on Oceans and International Environment, Senate Foreign Relations Committee, May 3, 1972, PPRT, LC; Transcript, Statement of Russell Train before Subcommittee on International Organization and Movements, House Foreign Affairs Committee, March 16, 1972, PPRT, LC.

54. Memo, Russell Train to John Whitaker, March 10, 1972, PPRT, LC; Memo, Russell Train to John Whitaker, March 20, 1972, Folder "John Ehrlichman, March, 1972," Box 93, H. R. Haldeman Files, WHSF, RNPMP; Memo, Russell Train to John Ehrlichman, April 6, 1972, PPRT, LC.

55. Memo, William Lestor to John Whitaker, April 10, 1972, Folder "CEQ, 1971–72, 1 of 3," Box 126, John Whitaker Files, WHCF, RNPMP; Letter, Russell Train to Jack Davis, May 2, 1972, PPRT, LC.

56. Memo, Acting Secretary of State to Russell Train, June 5, 1972, PPRT, LC; Instructions for Heads of United States Delegations to International Conferences, undated, PPRT, LC.

57. R. Stephen Berry, "What Happened at Stockholm," *Science and Public Affairs* 28, no. 7 (September 1972): 16–58; Letter, George Bush to Russell Train, October 25, 1972, PPRT, LC; Letter, Russell Train to George Bush, October 31, 1972, PPRT, LC.

58. Interview, Author with Russell Train, January 8, 2004; Train quoted in *San Francisco Chronicle*, June 8, 1972, 11.

59. Interview, Author with Christian Herter, January 17, 1991; Interview, Author with Russell Train, January 8, 2004.

60. Fact Sheet, U.N. Conference on the Human Environment, June 20, 1972, PPRT, LC; Achievements of the Stockholm Conference, undated, PPRT, LC; Memo, Russell Train to John Whitaker, November 28, 1972, PPRT, LC; Memorandum for the Files, Russell Train, November 2, 1972, PPRT, LC.

61. Memo, Russell Train to Richard Nixon, June 19, 1972, PPRT, LC; Transcript, Press Conference, Russell Train, Howard Baker and Robert White, June 20, 1972, PPRT, LC; Statement by the President on the U.N. Conference on the Human Environment, June 20, 1972, PPRT, LC.

62. Letter, Maurice Strong to Russell Train, July 28, 1972, PPRT, LC; Letter, Claiborne Pell to Russell Train, July 20, 1972, PPRT, LC; Letter, N. B. Livermore to Richard Nixon, August 8, 1972, PPRT, LC; *Congressional Record*, 92nd Cong., 2nd sess., vol. 118, pt. 6 (June 26, 1972): H6099–H6103.

63. Train, *Politics, Pollution, and Pandas*, 139–140; *New York Times*, August 30, 1972, 24.

64. P.L. 92–516; Flippen, *Nixon and the Environment*, 134, 178; Train, *Politics, Pollution, and Pandas*, 98.

65. Report of the Department of State on the Convention on International Trade in Endangered Species of Wild Fauna and Flora, April 5, 1973, Folder "EX NR Natural Resources, 3 of 3, 1973–74," Box 1, Natural Resources Files, WHCF, RNPMP; Press Release, CITES, April 13, 1973, Folder "EX NR Natural Resources, 3 of 3, 1973–74," Box 1, Natural Resources Files, WHCF, RNPMP; Dunlap, *Saving America's Wildlife*, 152–153; Train, *Politics, Pollution, and Pandas*, 144–145.

66. Memorandum for the Files, January 12, 1971, PPRT, LC; *Arizona Daily Star*, November 9, 1971, 1; Russell Train, "A New Approach to International Environmental Cooperation: The NATO Committee on the Challenges of Modern Society," *Kansas Law Review* 22, no. 21 (Winter 1974): 172–174.

67. Memorandum of Conversation, Russell Train, June 30, 1971, PPRT, LC; Letter, Richard Nixon to Russell Train, January 26, 1972, PPRT, LC; Memo, Russell Train to Henry Kissinger, April 13, 1972, PPRT, LC.

68. Memorandum for the Files, Russell Train, January 12, 1971, PPRT, LC.

69. The Soviets boycotted Stockholm because of the exclusion of East Germany; Summary of Environmental Problems in the Soviet Union, undated, 1971, PPRT, LC.

70. Memorandum for the Record, Gordon MacDonald, May 5, 1970, PPRT, LC; Memo, Russell Train to Henry Kissinger, October 28, 1971, PPRT, LC; Memo, Russell Train to Morton Hillenbrand, January 16, 1971, PPRT, LC; Letter, Russell Train to Fitzhugh Green, November 15, 1971, PPRT, LC; Memorandum for the Files, November 12, 1971, PPRT, LC.

71. Memo, Russell Train to John Ehrlichman, July 14, 1972, PPRT, LC; Memorandum for the Files, Russell Train, July 28, 1972, PPRT, LC; Memo, Jack Perry to Russell Train, July 27, 1972, PPRT, LC; Interview, Author with Russell Train, January 8, 2004; Train, *Politics, Pollution, and Pandas,* 127.

72. Cooperation in Environmental Protection: An Agreement between the United States of America and the Union of Soviet Socialist Republics, May 23, 1972, PPRT, LC; "Scientific Agreements of Moscow Summit," *BioScience* 22, no. 7 (July 1972): 424; Memo, Edwin Coate to John Whitaker, May 20, 1972, Folder "CEQ, 1972–73, 1 of 2," Box 43, John Whitaker Files, WHCF, RNPMP; Train quoted in *Washington Post,* May 24, 1972, 11.

73. Letter, Richard Nixon to Russell Train, August 4, 1972, PPRT, LC; Memo, Russell Train to John Whitaker, September 1, 1972, PPRT, LC; Memo, Lee Talbot to Russell Train, July 27, 1972, PPRT, LC; Memo, Russell Train to John Ehrlichman, September 7, 1972, PPRT, LC.

74. Memorandum for the Files, Russell Train, September 11, 1972, PPRT, LC.

75. Transcript, Remarks of the Honorable Russell E. Train, Chairman, U.S. Delegation, September 18, 1972, PPRT, LC; *New York Times,* September 22, 1972, 1; Itinerary, First Session of the US-USSR Joint Committee on Cooperation in the Field of Environmental Protection, September 18–22, 1972, PPRT, LC.

76. Interview, Author with Russell Train, January 8, 2004; Letter, Russell Train to Leontii Mirodonov, October 18, 1972, PPRT, LC; Letter, Russell Train to Yuri Izrael, October 18, 1972, PPRT, LC; *Washington Post,* October 5, 1972, 21.

77. *Baltimore Sun,* October 10, 1972, 10; Memo, Russell Train to Henry Kissinger, October 6, 1972, PPRT, LC.

78. Memorandum for the Files, Russell Train, February 1, 1973, PPRT, LC; Press Release, Soviets-US Hold Joint Press Conference, July 5, 2004, PPRT, LC; Letter, George Milias to Russell Train, July 5, 1973, PPRT, LC; Letter, Russell Train to Robert Long, August 6, 1973, PPRT, LC; Nixon quoted in Train, *Politics, Pollution, and Pandas,* 129.

79. Jonathan Aitken, *Nixon: A Life* (Washington, D.C.: Regnery Publishing, 1993), 423.

80. Letter, Daniel Schorr to Ron Zeigler, May 7, 1972, PPRT, LC; Letter, Diane Sawyer to Russell Train, August 3, 1972, PPRT, LC; Interview, Author with Russell Train, January 8, 1972.

81. Council on Environmental Quality, *The 1973 Annual Report of the President's Council on Environmental Quality* (Washington, D.C.: U.S. Government Printing Office, 1973); *Washington Evening Star and Daily News,* August 7, 1972, 1; August 11, 1972, 17; *Washington Post,* August 8, 1972, 1; August 10, 1972, 2; Joseph Browder, "Decision-Making in the White House," in Rathlesberger, *Nixon and the Environment,* 266–267; Interview, Author with William Reilly, June 26, 1998; *Hartford Times,* August 11, 1972, 11.

82. Richard N. L. Andrews, *Managing the Environment, Managing Ourselves: A History of American Environmental Policy* (New Haven, Conn.: Yale University Press, 1999), 243.

83. *Washington Post,* November 4, 1971, 1, 14; *International Herald Tribune,* November 6, 1971, 4; Memo, Russell Train to Richard Fairbanks, January 19, 1973, PPRT, LC; Memo, Russell Train to Richard Fairbanks, January 26, 1972, PPRT, LC.

84. *Public Papers of the Presidents: Richard Nixon, 1971* (Washington, D.C.: U.S. Government Printing Office, 1969–1974), 146–147; Report of the National Industrial Pollution Control Council, February 10, 1971, Folder "EX FG 278, NIPCC, January 1, 1971–1972," Box 1, NIPCC Files, WHCF, RNPMP; Hays, *Beauty, Health, and Permanence*, 214–215; Memo, John Ehrlichman to Edward David, May 16, 1972, Folder "White House—Environment, 1972," Box 9, Edward David Files, WHCF, RNPMP; Memo, Leonard Laster to Kenneth Cole, January 5, 1973, Folder "White House Life Sciences, 1973," Box 12, Edward David Files, WHCF, RNPMP.

85. Cahn quoted in *New York Times*, September 6, 1972, 12; Interview, Author with Russell Train, January 8, 2004; *Los Angeles Times*, September 8, 1972, D3; Memo, Richard Fairbanks to Neal Ball, September 1, 1972, Folder "CEQ, 1972, 1 of 2," Box 126, John Whitaker Files, WHCF, RNPMP; Memo, Daniel Kingsley to Richard Nixon, September 22, 1972, Folder "Presidential Handwriting, September 15–30, 1972," Box 18, President's Office Files, WHSF, RNPMP.

86. Richard Nathan, *The Plot that Failed: Nixon and the Administrative Presidency* (New York, N.Y.: John Wiley and Sons, 1975), 61–65; Interview, Author with Russell Train, January 8, 2004.

87. Memo, Earl Butz to Russell Train, William Ruckelshaus, and Rogers Morton, March 12, 1972, Folder "EX FG 999–28, DNR, 1971–1973," Box 14, Proposed Departments, WHCF, RNPMP; Interview, Author with Russell Train, January 8, 2004; Train quoted in *Environmental Quality Magazine* (June 1973): 21.

88. Interview, Author with Russell Train, July 8, 1998; Train quoted in Interview, Author with Richard Fairbanks, March 24, 1998.

89. *New York Times*, January 1, 1972, 36; January 24, 1972, 10; March 3, 1972, 10; April 26, 1972, 65; May 1, 1972, 26; June 20, 1972, 51; Train, *Politics, Pollution, and Pandas*, 163; Frederick R. Anderson and Robert H. Daniels, *NEPA in the Courts* (Baltimore, Md.: Johns Hopkins University Press, 1973), x; Walter Rosenbaum, *The Politics of Environmental Concern* (New York, N.Y.: Praeger, 1974), 270.

90. Memo, Russell Train to John Ehrlichman, December 7, 1971, PPRT, LC; Train quoted in *Washington Evening Star and Daily News*, June 14, 1973, 1.

91. Memo, John Whitaker to William Ruckelshaus, March 1, 1972, Folder "1972 Campaign, 3 of 3," Box 33, John Whitaker Files, WHCF, RNPMP; Memo, John Whitaker to Russell Train, March 3, 1972, Folder "1972 Campaign, I, 2 of 5," Box 123, John Whitaker Files, WHCF, RNPMP.

92. Interview, Author with Russell Train, January 8, 2004.

93. Transcript, Remarks of the Honorable Russell E. Train before the Platform Committee of the Republican National Convention, August 17, 1972, Folder "1972 Campaign, I, 1 of 2," Box 124, John Whitaker Files, WHCF, RNPMP; Train, *Politics, Pollution, and Pandas*, 111.

94. See Donald Johnson and Kirk Porter, *National Party Platforms* (Urbana: University of Illinois Press, 1973); *Washington Post*, November 8, 1972, 1; *New York Times*, November 8, 1972, 1; Ambrose, *Nixon: The Triumph of a Politician*, 651–652.

95. Interview, Author with Russell Train, January 8, 2004; Interview, Author with Aileen Train, January 9, 2004, Aileen quoted in Train, *Politics, Pollution, and Pandas*, 112.

96. Memorandum for the Files, Russell Train, January 12, 1973, PPRT, LC; see Stanley Kutler, *Abuse of Power* (New York, N.Y.: Free Press, 1997).

97. Memorandum for the Files, Russell Train, May 4, 1973, PPRT, LC; Train, *Politics, Pollution, and Pandas*, 112–114.

98. Letter, Russell Train to Al Haig, September 8, 1973, PPRT, LC.

99. Russell Train, *A Memoir* (Washington, D.C.: self-published, 2000), 196.

CHAPTER 7

1. See Fred Emery, *The Corruption of American Politics and the Fall of Richard Nixon* (New York, N.Y.: Touchstone, 1994); Keith Olson, *The Presidential Scandal that Shocked America* (Lawrence: University of Kansas Press, 2003); Monica Crowley, *Nixon in Winter* (New York, N.Y.: Random House, 1998), and John Ehrlichman, *Witness to Power: The Nixon Years* (New York, N.Y.: Pocket Books, 1982).

2. Interview, Author with William Ruckelshaus, August 25, 2004.

3. Quoted in *Environmental Action* (August 4, 1973): 12.

4. Press Release, Office of White House Press Secretary, July 26, 1973, PPRT, LC; Interview, Author with Russell Train, January 8, 2004.

5. *Suburban and Wayne Times,* August 2, 1973, 13.

6. Quoted in *Time* (April 6, 1973): 24.

7. *Wall Street Journal,* July 25, 1973, 1.

8. *Congressional Record,* 93rd Cong., 1st sess., vol. 119, pt. 27 (September 5, 1973), S28469; Russell Train, *A Memoir* (Washington, D.C.: self-published, 2000), 198.

9. Interview, Author with Russell Train, January 8, 2004.

10. Ibid.; *Washington Post,* July 25, 1973, 2; *National Observer,* August 25, 1973, 12; *Phoenix Gazette,* September 11, 1973, 6; *Wall Street Journal,* July 27, 1973, 1; *Tulsa Daily World,* September 23, 1973, 5; Nader quoted in *Newsweek* (August 6, 1973): 31.

11. Quoted in *Christian Science Monitor,* July 28, 1973, 3.

12. Ibid.; U.S. Congress, Senate Committee on Public Works, *Hearings on the Nomination of Russell E. Train to be Administrator of the Environmental Protection Agency,* 93rd Cong., 1st sess., August 1, 1973, 3.

13. Memo, Roger Strelow to Russell Train, July 23, 1973, PPRT, LC; Letter, Russell Train to William Scott, August 6, 1973, PPRT, LC; Letter, Russell Train to Jennings Randolph, August 21, 1973, PPRT, LC; Letter, Russell Train to William Harsha, August 28, 1973, PPRT, LC; Notebook of Congratulatory Letters, September, 1973, PPRT, LC; *Salt Lake Tribune,* September 11, 1973, 4; *Rocky Mountain News,* August 14, 1973, 43; *Seattle Times,* August 9, 1973, 13.

14. John Quarles, *Cleaning Up America: An Insider's View of the Environmental Protection Agency* (Boston, Mass.: Houghton Mifflin, 1976), 119, 199.

15. Nixon quoted in Interview, Author with Russell Train, January 8, 2004; Letter, Richard Nixon to Russell Train, September 18, 1973, PPRT, LC.

16. Memo, John Ehrlichman to Peter Flanigan, August 25, 1971, Folder "John Ehrlichman, August, 1971," Box 83, H. R. Haldeman Files, WHSF, RNPMP; Memo, John Whitaker to Peter Flanigan, September 10, 1971, Folder "EPA, 1951, 3 of 4," Box 64, John Whitaker Files, WHCF, RNPMP.

17. Nixon quoted in Memorandum for the President's File, July 26, 1973, Folder "Meetings File, Beginning July 22, 1973," Box 92, President's Office Files, WHSF, RNPMP.

18. *Washington Post,* July 19, 1973, 1; Memo, Roger Strelow to Russell Train, July 20, 1973, PPRT, LC.

19. Marc Landy, Marc Roberts, and Stephen Thomas, *The Environmental Protection Agency: Asking the Wrong Questions; From Nixon to Clinton* (New York, N.Y.: Oxford University Press, 1994), 37; Domestic Council Study Memorandum no. 15, June 26, 1971, Folder "Ex FG 298, EPA, January 1, 1971–1972, 1of 3," Box 1, EPA Files, WHCF, RNPMP; Memo, Edward David to Members, Domestic Council Committee on Quality of Life, June 28, 1971, Folder "Quality of Life, 1971 (1970–72), 3 of 4," Box 96, John Whitaker Files, WHCF, RNPMP; Interview, Author with Russell Train, July 8, 1998; *New York Times,* November 8, 1971, 28; November 18, 1971, 22.

20. Memo, Roger Strelow to Russell Train, July 16, 1973, PPRT, LC.

21. Interview, Author with Laurence Moss, March 25, 1998.

22. Interview, Author with William Ruckelshaus, August 25, 2004.

23. *Washington Post*, July 25, 1973, 2; Russell E. Train, *Politics, Pollution, and Pandas* (Washington, D.C.: Island Press, 2003), 159–160.

24. See Stephen Ambrose, *Nixon: Ruin and Recovery, 1972-1990* (New York, N.Y.: Simon and Schuster, 1991), 229–262.

25. Ruckelshaus served as FBI director for only three months before Nixon named him Deputy Attorney General.

26. Interview, Author with Russell Train, January 8, 2004; Train, *Politics, Pollution, and Pandas*, 152.

27. Memorandum for the Files, Russell Train, September 8, 1973, PPRT, LC; Train, *Politics, Pollution, and Pandas*, 163; Train, *A Memoir*, 269.

28. Quoted in *Wall Street Journal*, March 22, 1974, 19.

29. U.S. Congress, Senate Committee on Public Works, *Hearings on the Compliance with Title II (Auto Emission Standards) of the Clean Air Act*, 93rd Cong., 1st sess., November 6, 1973, 431–436.

30. *Public Papers of the Presidents: Richard Nixon, 1973* (Washington, D.C.: U.S. Government Printing Office, 1969–1974), 916–922; *Nation* (January 5, 1974): 11–16; *New Republic* (December 29, 1973): 5–6; *Washington Post*, March 4, 1973, 4.

31. *Newsweek* (March 18, 1974): 99; *Business Week* (March 16, 1974): 27; *New York Times*, March 11, 1974, 48; March 12, 1974, 16.

32. By this time, Nixon had replaced Love for not being a forceful leader.

33. Nixon quoted in *New York Times*, March 11, 1974, 1; *Motor* (May 1974): 6; *Wall Street Journal*, March 19, 1974, 22.

34. *Washington Star*, March 11, 1974, 1; March 22, 1974, 1; March 23, 1974, 23; *Wall Street Journal*, March 15, 1974, 6; Fact Sheet, Clean Air Act Proposal, March, 1974, Folder "EX 9–1, Air Pollution, September, 1973–, 2 of 2," Box 31, Health Files, WHCF, RNPMP.

35. P.L. 93–319, 88 Stat. 246; Train, *Politics, Pollution, and Pandas*, 173.

36. *Washington Post*, March 23, 1974, 1.

37. Letter, unsigned, to Russell Train, November 11, 1973, PPRT, LC; Letter, Gehring Dietrick to Russell Train, February 2, 1974, PPRT, LC.

38. *Washington Post*, March 31, 1974, 21.

39. Russell Train, "An Environmental Sell-Out Will Not Turn Energy Faucets on Full," in *American Lung Association Bulletin* (March 1974): 2–3.

40. Quoted in *Sewanee News* (March 1974): 5; see also *New York Times*, September 13, 1973, 40; *Washington Post*, April 15, 1974, 2; *U.S. News and World Report* (March 11, 1974): 39–43.

41. Quoted in *Cleveland Plain Dealer*, June 25, 1974, 15; *New York Times*, June 27, 1974, 13.

42. Quarles, *Cleaning Up America*, 193–211; Train, *Politics, Pollution, and Pandas*, 175–176; *New York Times*, April 19, 1974, 23.

43. *Montana Standard*, July 10, 1974, 3; *Washington Post*, July 9, 1974, 3; *Washington Star*, July 21, 1974, 17; *New York Times*, July 14, 1974, 20; *New Republic* (July 20, 1974): 7.

44. *New Republic* (March 9, 1974): 3; Scott quoted in *Congressional Record*, 93rd Cong., 1st sess., vol. 119, pt. 27 (September 5, 1973), S28469; Transcript, Interview with Russell Train, WTTG-TV, Washington, D.C., January 14, 1974, PPRT, LC.

45. Comprising the Seven Sisters were Standard Oil of New Jersey, Royal Dutch Shell group, Texaco, Standard Oil of California, Socony Mobil, Gulf, and British Petroleum; Joseph M. Petulla, *American Environmental History* (San Francisco, Calif.: Boyd and Fraser, 1977), 345; *New York Times*, March 7, 1974, 1; March 22, 1974, 62; *Cleveland Press*, September 28, 1973, 4.

46. Letter, Patrick F. Noonan to Russell Train, April 22, 1974, PPRT, LC; *American Forests* (May 1974): 7.

47. Shell spokesman quoted in *New York Times*, August 3, 1974; Train quoted in *Washington Post*, June 12, 1974, 13; *Wall Street Journal*, April 25, 1974, 4.

48. See Brent Walth, *Fire at Eden's Gate: Tom McCall and the Oregon Story* (Portland: Oregon Historical Society Press, 1994; J. Brooks Flippen, *Nixon and the Environment* (Albuquerque: University of New Mexico Press, 2000), 217; Train *Politics, Pollution, and Pandas*, 166; *Eugene Register Guard*, February 7, 1973, 3.

49. See J. Brooks Flippen, "Containing the Urban Sprawl: The Nixon Administration's Land Use Policy," *Presidential Studies Quarterly* 26, no. 1 (Winter 1996): 197–207; *Washington Post*, March 26, 1974, 17; June 11, 1974, 13; *New York Times*, July 9, 1974, 16; July 12, 1974, 32.

50. Quoted in *New York Times*, April 7, 1974, 4.

51. Ibid., April 11, 1974, 4; April 3, 1974, 12.

52. Train, *A Memoir*, 206–207.

53. Letter, Roger Egeberg to Russell Train, November 29, 1973, PPRT, LC.

54. *Martinsburg Journal*, April 8, 1974, 1; Diary of Russell Train, 1976 Journal, January 23, 1976, PPRT, LC.

55. Uncited news clippings, PPRT, LC; Interview, Author with Aileen Train, January 9, 2004; Letter, Russell Train to Lady Douglas Hamilton, undated, PPRT, LC.

56. Letter, Russell Train to Robert Long, August 6, 1973, PPRT, LC; List of Tentative Dates, US-USSR Environmental Program, July 26, 1973, PPRT, LC; Memorandum for the Files, November 14, 1973, PPRT, LC; Letter, Russell Train to Yuri Izrael, July 23, 1973, PPRT, LC; Letter, Russell Train to A. O. Kudryavtsov, August 3, 1973, PPRT, LC; Unsigned Tribute to Russell Train, September 20, 1994, PPRT, LC.

57. Press Release, Embassy of the United States of America, May 5, 1974, PPRT, LC; Itinerary, Visit of Russell Train, Bonn, West Germany, May 9–10, 1974, PPRT, LC; Agreement between the Government of the United States of America and the Government of Federal Republic of Germany on Cooperation in the Field of Environmental Protection, PPRT, LC; Letter, E. Vernede to Russell Train, May 24, 1974, PPRT, LC; Letter, Sadanori Yamanaka to Russell Train, September 26, 1974, PPRT, LC.

58. Train quoted in Letter, Michael Raoul-Duval to Russell Train, August 6, 1974, PPRT, LC; *Laconia (N.H.) News*, July 18, 1974, 22; Terence Kehoe, *Cleaning Up the Great Lakes: From Cooperation to Confrontation* (DeKalb: Northern Illinois University Press, 1997), 164–165; Memo, Theodore L. Eliot to Henry Kissinger, February 2, 1973, Folder "EX NR 7–1, Projects 3 of 3, 1973–1974," Box 17, Natural Resources Files, WHCF, RNPMP.

59. Transcript, Interview with Russell Train, WTTG-TV, January 14, 1974, PPRT, LC; Letter, David Bardin to Russell Train, June 21, 1974, PPRT, LC.

60. *Train v. City of New York*, 420 U.S. 35 (1975); Interview, Author with Russell Train, January 9, 2004; Harvey Lieber, *Federalism and Clean Waters: The 1972 Water Pollution Control Act* (Lexington, Mass.: D.C. Heath and Co., 1975), 118, 120–121, 123.

61. *United States of America v. Richard Nixon*, 418 U.S. 683 (1974).

62. Ambrose, *Nixon: Ruin and Recovery*, 394–399; Richard Nixon, *RN: The Memoirs of Richard Nixon* (New York, N.Y.: Grosset and Dunlap, 1978); Henry Kissinger, *Years of Upheaval* (Boston, Mass.: Little, Brown, 1982), 1207–1209.

63. Memo, Richard Nixon to Russell Train, March 22, 1974, Folder "EX FG 298, EPA, January 1, 1974–May 30, 1974," Box 2, EPA Files, WHCF, RNPMP.

64. Memorandum for the Files, July 15, 1974, Folder "Meetings File, Beginning July 7, 1974 ," Box 94, WHSF, RNPMP; Opinion Column, Edward Morgan, ABC News, Washington, August 8, 1974, PPRT, LC.

65. *Economist* (March 30, 1974): 45.

66. Jackson quoted in *BioScience* 24, no. 8 (August 1974): 470–471; Martin quoted in *New York Times,* July 2, 1974, 23.

67. Train, *A Memoir,* 271; Train quoted in undated, 1974 Associated Press article, PPRT, LC.

68. Robert L. Sansom, *The New American Dream Machine: Towards a Simpler Lifestyle in an Environmental Age* (Garden City, N.J.: Anchor Press, 1976), 25.

69. Harlow quoted in Memorandum for the Files, Russell Train, October 17, 1974, PPRT, LC; Interview, Author with Russell Train, January 9, 2004.

CHAPTER 8

1. Albert quoted in Congressional Quarterly, *Gerald Ford: The Man and His Record* (Washington, D.C.: Congressional Quarterly, Inc., 1974), 31.

2. Bob Woodward, *Five Presidents and the Legacy of Watergate* (New York, N.Y.: Simon and Schuster, 1999), 29; Derwinski and Ford quoted in ibid., 28 and 67, respectively.

3. Ford and Train quoted in Congressional Quarterly, *Gerald Ford,* 7 and 27, respectively; Interview, Author with Russell Train, January 9, 2004; see *Nelson Rockefeller: Passionate Millionaire* (Videocassette, ABC News Productions, 1998).

4. Letter, Russell Train to Gerald Ford, August 9, 1974, PPRT, LC.

5. Rumsfeld quoted in Train, Russell E. Train, *Politics, Pollution, and Pandas* (Washington, D.C.: Island Press, 2003), 182.

6. Memo, Russell Train to Gerald Ford, August 13, 1974, PPRT, LC.

7. Gerald Ford, *A Time to Heal: The Autobiography of Gerald R. Ford* (New York, N.Y.: Harper and Row, 1979), 130–131.

8. Interview, Author with Russell Train, January 9, 2004; *Time* (October 21, 1974): 19.

9. *Washington Post,* September 3, 1974, 12; Ford quoted in Memorandum for the Files, October 17, 1974, Russell Train, PPRT, LC.

10. Memorandum for the Files, October 17, 1974, Russell Train, PPRT, LC.

11. Ibid.; Interview, Author with Russell Train, January 9, 2004.

12. Memorandum for the Files, undated, Russell Train, PPRT, LC.

13. Ibid.

14. Brower quoted in *Time* (October 21, 1974): 21.

15. Memorandum for the Files, October 23, 1974, Russell Train, PPRT, LC.

16. *Washington Post,* September 14, 1974, 3; Memorandum for the Files, February 24, 1975, Russell Train, PPRT, LC; Memorandum for the Files, January 8, 1975, Russell Train, PPRT, LC; Letter, Nelson Rockefeller to Russell Train, April 8, 1975, PPRT, LC; Memorandum for the Files, February 18, 1975, Russell Train, PPRT, LC.

17. Train quoted in *Chicago Tribune,* October 18, 1974, 21; *Easton Star Democrat,* September 6, 1974, 12; and *Monterey Peninsula Herald,* October 31, 1974, 4.

18. Quoted in *Athens Banner Herald*, September 6, 1974, 1; *Washington Star*, September 13, 1974, 2.

19. Ford, *A Time to Heal*, 125, 228; quoted in Congressional Quarterly, *Gerald Ford*, 7.

20. Hal K. Rothman, *The Greening of a Nation?: Environmentalism in the United States since 1945* (Fort Worth, Tex.: Harcourt Brace, 1998), 132–133.

21. Michael Schaller and George Rising, *The Republican Ascendancy: American Politics, 1968–2001* (Wheeling, Ill.: Harlan Davidson, 2002), 65.

22. John Micklethwait and Adrian Wooldridge, *The Right Nation: Conservative Power in America* (New York, N.Y.: Penguin, 2004), 77.

23. H. W. Brands, *The Strange Death of American Liberalism* (New Haven, Conn.: Yale University Press, 2001), 113.

24. Viguerie quoted in Lee Edwards, *The Conservative Revolution: The Movement that Remade America* (New York, N.Y.: Free Press, 1999), 183.

25. Quoted in ibid., 179.

26. Betty Ford, *The Times of My Life* (New York, N.Y.: Harper and Row, 1978), 210; LaHaye quoted in Schaller and Rising, *Republican Ascendancy*, 63.

27. Ruth Murray Brown, *For a Christian Nation: A History of the Religious Right* (Amherst, N.Y.: Prometheus Books, 2002), 133; see also Irving Kristol, *Neo-Conservativism: The Autobiography of an Idea; Selected Essays* (New York, N.Y.: Free Press, 1995).

28. Quoted in Schaller and Rising, *Republican Ascendancy*, 63; Congressional Quarterly, *Gerald Ford*, 16.

29. Diary of Russell Train, 1976 Journal, September 11, 1976, PPRT, LC.

30. *San Jose Mercury*, October 31, 1974, 17; *Washington Post*, November 5, 1974, 3; Memorandum for the Files, February 10, 1975, Russell Train, PPRT, LC.

31. Memorandum for the Files, December 2, 1974, PPRT, LC.

32. Memorandum for the Files, January 7, 1975, Russell Train, PPRT, LC.

33. Ibid.

34. Transcript, Interview, Russell Train with WETA-TV, Evening Edition, December 23, 1975, PPRT, LC; *U.S. News and World Report* (August 18, 1975): 21; *Washington Post*, May 31, 1975, 1; *New York Times*, May 31, 1975, 1.

35. Memorandum for the Files, March 3, 1975, Russell Train, PPRT, LC; *Washington Post*, December 25, 1974, 20; *New York Times*, December 25, 1974, 1.

36. Quoted in Train, *Politics, Pollution, and Pandas*, 171.

37. Goldwater quoted in *Arizona Republic*, August 1, 1976, 4; Letter, Russell Train to Barry Goldwater, August 18, 1976, PPRT, LC.

38. Letter, Barry Goldwater to Russell Train, August 25, 1976, PPRT, LC; Letter, Russell Train to Barry Goldwater, undated, PPRT, LC.

39. Memorandum for the Files, July 7, 1975, Russell Train, PPRT, LC; *New York Times*, July 9, 1975, 23; Ford quoted in Mary H. Cooper, "Environmental Movement at 25," *Congressional Quarterly Researcher* (March 31, 1995): 418.

40. Train quoted in *Washington Star*, September 9, 1975, 13.

41. Memo, James Connor to Cabinet Members, April 14, 1975, PPRT, LC; Memo, Alvin Alm to Russell Train, March 26, 1975, PPRT, LC; Memo, Alvin Alm to Russell Train, February 24, 1975, PPRT, LC; *Washington Post*, October 4, 1974, 12; *Oregonian*, November 2, 1974, 19; *Cleveland Plain Dealer*, January 1, 1975, B2.

42. Diary of Russell Train, 1976 Journal, March 4, 1976, PPRT, LC; Train, *Politics, Pollution, and Pandas*, 212.

43. Diary of Russell Train, 1976 Journal, June 1, 1976, PPRT, LC; Baker quoted on June 8, 1976.

44. Ibid., June 11, 1976, PPRT, LC.

45. Ibid., June 15, 1976, PPRT, LC.

46. Ibid., July 19, 1976, PPRT, LC; Train, *Politics, Pollution, and Pandas*, 218.

47. Diary of Russell Train, 1976 Journal, September 3, 1976, PPRT, LC; Surface Mining Control and Reclamation Act; Memorandum for the Files, May 28, 1995, Russell Train, PPRT, LC; Ford quoted in *A Time to Heal*, 226; see also Council on Environmental Quality, *Coal Surface Mining and Reclamation: An Environmental and Economic Assessment of Alternatives* (Washington, D.C.: U.S. Government Printing Office, 1973); John Whitaker, *Striking a Balance: Environment and Natural Resources Policy in the Nixon-Ford Years* (Washington, D.C.: American Enterprise Institute, 1976), 173–183.

48. Train, *Politics, Pollution, and Pandas*, 223; Train quoted in the *Energy Daily* 5, no. 35 (February 18, 1977): 1.

49. The tankers were the *Argo Merchant* and the *Olympic Games;* Whitaker, *Striking a Balance*, 208–218; *Washington Star,* December 28, 1976, 1; December 29, 1976, 1; *Washington Post,* December 24, 1976, 16; December 22, 1976, 1, 4; Transcript, Interview with Russell Train, WTOP Radio, CBS Network, December 23, 1976, PPRT, LC; Diary of Russell Train, 1976 Journal, December 20, 1976, December 21, 1976, PPRT, LC.

50. Transcript, Interview, Russell Train, WMAL-TV, August 17, 1976, PPRT, LC; Transcript, Interview, Elmo Zumwalt, WTOP-TV, August 15, 1976, PPRT, LC; Transcript, Interview, Russell Train, WRC-TV, August 17, 1976, PPRT, LC; *Baltimore Sun,* October 2, 1976, 15; *The Banner,* September 2, 1976, 1; *Washington Star,* August 21, 1976, 19; *Washington Post,* August 18, 1976, 1.

51. Transcript, Interview, Russell Train, WTOP-TV, November 19, 1975, PPRT, LC; Interview, Russell Train, CBS News, December 23, 1975, PPRT, LC.

52. *Time* (October 25, 1976): 21; *Oregon Journal,* September 15, 1976; U.S. Congress, Senate Committee on Commerce, *Report of the Senate Committee on Commerce on S.3149, Toxic Substances Control Act, March 16, 1976* (Washington, D.C.: U.S. Government Printing Office, 1976), 3.

53. Train, *Politics, Pollution, and Pandas*, 231; see U.S. Department of Agriculture, Food Safety and Quality Service, *Report on the PCB Incidence in the Western United States* (Washington, D.C.: U.S. Government Printing Office, 1980), and *EDF v. EPA* November 3, 1978 (12 ERC 1353); Samuel Hays, *Beauty, Health, and Permanence: Environmental Politics in the United States, 1955–1985* (New York, N.Y.: Cambridge University Press, 1987), 193–195.

54. Transcript, Interview with Russell Train, WMAL-TV, ABC Network, August 4, 1975, PPRT, LC; *Atlanta Journal,* January 5, 1977, 14; *Houston Post,* August 23, 1975, 19; *Des Moines Register,* July 17, 1975, 5.

55. Train quoted in *New York Times,* December 24, 1976, 3; Memorandum for the Files, Russell Train, May 6, 1975, PPRT, LC.

56. Memo, Russell Train to James Connor, May 2, 1975, PPRT, LC.

57. Safe Drinking Water Act of 1974; *Philadelphia Inquirer,* April 19, 1975, 3; *Charlotte Observer,* April 19, 1975, 3; Train, *Politics, Pollution, and Pandas,* 192.

58. See Diary of Russell Train, 1976 Journal, Trip to USSR, Japan and China, PPRT, LC; *Los Angeles Times,* May 17, 1976, 7; Memorandum for the Files, October 29, 1976, PPRT, LC; Transcript, Remarks of Russell Train to CCMS 1975 Fall Plenary Meeting, October 15, 1975, PPRT, LC; Department of State Newsletter, December 1974, 5, PPRT, LC; *Kayhan International* 18, no. 5797 (November 11, 1976): 7.

59. Podgorny quoted in Memorandum for the Files, Russell Train, November 29, 1976, PPRT, LC;

Memorandum for the Files, Russell Train, December 17, 1974, PPRT, LC; *EPA Journal* 1, no. 8 (December 1975): 15–16.

60. Diary of Russell Train, 1976 Journal, March 11, 1976, May 7, 1976, PPRT, LC; Letter, Robert Byrd to Russell Train, December, 1975, PPRT, LC; Letter, Lady Bird Johnson to Russell Train, October 11, 1974, PPRT, LC.

61. Letter, Robert Michel to Russell Train, December 9, 1975, PPRT, LC; Letter, Caspar Weinberger to Russell Train, undated, PPRT, LC: Letter, Robert Barker to Russell Train, April 14, 1975, PPRT, LC; Letter, Maurice Stans to Russell Train, May 15, 1975, PPRT, LC.

62. Diary of Russell Train, 1976 Journal, May 4, 1976, PPRT, LC; quoted in *Star Democrat,* February 21, 1975, 3.

63. Memorandum for the Files, Russell Train, August 27, 1975, PPRT, LC; Diary of Russell Train, 1976 Journal, August 30, 1976, February 26, 1976, March 4, 1976, January 1, 1976, PPRT, LC.

64. Diary of Russell Train, 1976 Journal, June 22, 1976, PPRT, LC; *Washington Post,* October 24, 1976, B1; *Washington Star,* November 2, 1976, 14; Interview, Author with Russell Train, January 9, 2004.

65. Train quoted in *Sanford (Maine) Tribune,* March 12, 1976, 4; Letter, unsigned to Russell Train, October 12, 1976, PPRT, LC; Letter, unsigned to EPA, September 25, 1976, PPRT, LC;

66. Transcript, Interview with Russell Train, CBS Morning News, January 12, 1976, PPRT, LC; Diary of Russell Train, 1976 Journal, January 9, 1976, January 11, 1976, January 12, 1976, PPRT, LC.

67. Diary of Russell Train, 1976 Journal, June 25, 1976, May 18, 1976, PPRT, LC.

68. Rockefeller quoted in ibid., March 16, 1976, August 6, 1976, PPRT, LC.

69. Interview, Author with Russell Train, January 9, 2004.

70. Ibid.; Diary of Russell Train, 1976 Journal, January 22, 1976, August 9, 1976, August 19, 1976, PPRT, LC; Ford, *A Time to Heal,* 404.

71. Diary of Russell Train, 1976 Journal, June 15, 1976, July 20, 1976, August 16, 1976, PPRT, LC.

72. Diary of Russell Train, 1976 Journal, August 12, 1976, August 19, 1976, PPRT, LC.

73. Ibid.; Train, *Politics, Pollution, and Pandas,* 222.

74. Carter quoted in *Washington Post,* January 23, 1977, 21.

75. Interview, Author with Russell Train, January 9, 2004; Diary of Russell Train, 1976 Journal, September 15, 1976, July 12, 1976, PPRT, LC.

76. Carter quoted in Betty Glad, *In Search of the Great White House* (New York, N.Y.: Norton, 1980), 314.

77. See U.S. House of Representatives, Committee on House Administration, *The Presidential Campaign, 1976* (Washington, D.C.: U.S. Government Printing Office, 1978); Jimmy Carter, *Keeping Faith: Memoirs of a President* (New York, N.Y.: Bantam, 1982), 125.

78. Letter, Russell Train to Gerald Ford, January 11, 1977, PPRT, LC; Letter, Christopher Phillips to Russell Train, January 24, 1977, PPRT, LC; Letter, Oliver Houck to Russell Train, January 19, 1977, PPRT, LC; Letter, Herman Talmadge to Russell Train, January 27, 1977, PPRT, LC; Diary of Russell Train, 1976 Journal, November 23, 1976, November 30, 1976, PPRT, LC; Train, *Politics, Pollution, and Pandas,* 229.

79. *Los Angeles Times,* December 15, 1976, 2; *New York Times,* December 19, 1976, 29; January 23, 1976, 16; *Conservation Foundation Newsletter* (January 1977): 1–9.

80. Transcript, Interview, *Los Angeles Times* with Russell Train, January, 1977, PPRT, LC; Memorandum for the Files, January 10, 1988, PPRT, LC.

81. *Conservation Foundation Newsletter* (January 1977): 3.

CHAPTER 9

1. Diary of Russell Train, 1977 Journal, January 21, 1977, April 20, 1977, PPRT, LC.

2. Ibid., January 25, 1977, March 8, 1977, March 20, 1977, May 12, 1977, August 5, 1977, PPRT, LC.

`3. Ibid., May 25, 1977, June 1, 1977, June 2, 1977, PPRT, LC; Interview, Author with Russell Train, January 9, 2004.

4. Carroll W. Wilson et al., *Coal—A Bridge to the Future: A Report of the World Coal Study, WOCOL* (Cambridge, Mass.: Ballinger, 1980); Diary of Russell Train, 1980 Journal, May 12, 1980, PPRT, LC.

5. Diary of Russell Train, 1980 Journal, October 2, 1980, PPRT, LC; Interview, Author with Russell Train, January 9, 2004.

6. Diary of Russell Train, 1980 Journal, October 2, 1980, PPRT, LC; Interview, Author with Russell Train, January 9, 2004; Diary of Russell Train, 1979 Journal, September 24, 1979, PPRT, LC; Russell E. Train, *Politics, Pollution, and Pandas* (Washington, D.C.: Island Press, 2003), 258–259.

7. *New York Times*, December 20, 1980, 1, 24, 30; Letter, Patrick Noonan to Russell Train, December 31, 1980, PPRT, LC; David Schoenbrod, "Limits and Dangers of Environmental Mediation: A Review Essay," *New York University Law Review* 58, no.6 (December 1983): 1453–1476; Letter, Helen Fenske to Russell Train, January 1, 1981, PPRT, LC; Letter, Laurance Rockefeller to Russell Train, March 10, 1981, PPRT, LC.

8. Diary of Russell Train, 1977 Canadian Hunting Trip, February 12, 1977, PPRT, LC; Diary of Russell Train, 1977 Journal, February 12–27, 1977, May 1–11, 1977, PPRT, LC; Letter, Russell Train et al. to Prince Saud al Faisal, May 4, 1977, PPRT, LC.

9. Letter, S. Dillon Ripley to Russell Train, November 8, 1977, PPRT, LC.

10. Rockefeller quoted in Diary of Russell Train, 1977 Journal, December 16, 1977, PPRT, LC; Diary of Russell Train, 1978 Journal, January 11, 1978, PPRT, LC.

11. Diary of Russell Train, 1978 Journal, March 30, 1978, PPRT, LC.

12. Ibid., March 8, 1978, PPRT, LC; Letter, Russell Train to Gerald Ford, April 18, 1978, PPRT, LC.

13. Train, *Politics, Pollution, and Pandas*, 239; Diary of Russell Train, 1978 Journal, March 8, 1978, March 30, 1978, April 7, 1978, PPRT, LC.

14. Memorandum for the Files, Russell Train, March 1, 1979, PPRT, LC; Memorandum for the Files, Russell Train, September 19, 1978, PPRT, LC.

15. Diary of Russell Train, 1979 Journal, February 6, 1979, PPRT, LC; Train quoted in World Wildlife Fund–United States, *Annual Report, 1978* (Washington, D.C.: World Wildlife Fund, 1978), 1.

16. Diary of Russell Train, 1979 Journal, March 19, 1979, PPRT, LC; Interview, Author with Russell Train, January 9, 2004.

17. Transcript, Remarks of Honorable Russell E. Train, World Conference on Sea Turtles Convention, Department of State, Washington, D.C., November 26, 1979, PPRT, LC; Program, WCSTC, November 1979, PPRT, LC; Press Release, WCSTC, November 30, 1979, PPRT, LC.

18. Diary of Russell Train, 1978 Trip to Galapagos, Peru and Colombia, August 8, 1978, August 9, 1978, August 13, 1978, PPRT, LC; see Robert Blake, *Disraeli* (London, England: Eyre and Spottiswoode, 1966).

19. Activities Report, Russell Train to WWF Board, February 22, 1984; Diary of Russell Train, 1978 Trip to South America, August 6, 1978, PPRT, LC.

20. Diary of Russell Train, 1980 Journal, March 24, 1980, March 10, 1980, PPRT, LC; Train, *Politics, Pollution, and Pandas,* 283–284; Diary of Russell Train, 1978 Journal, April 16, 1978, PPRT, LC.

21. Diary of Russell Train, 1979 Journal, August 22, 1979, PPRT, LC; Diary of Russell Train, 1980 Journal, November 4, 1980, August 2, 1980, PPRT, LC.

22. Diary of Russell Train, 1979 Journal, September 3, 1979, May 2, 1979, August 2, 1979, PPRT, LC.

23. Ibid., February 28, 1979, February 22, 1979, PPRT, LC.

24. Bruce Mazlish and Edwin Diamond, *Jimmy Carter: A Character Portrait* (New York, N.Y.: Simon and Schuster, 1979), 231; Lawrence Shoup, *The Carter Presidency and Beyond* (Palo Alto, Calif.: Ramparts Press, 1980), 215.

25. Jeffrey K. Stine, "Environmental Policy during the Carter Presidency," in Gary M. Fink and Hugh Davis Graham, eds., *The Carter Presidency: Policy Choices in the Post–New Deal Era* (Lawrence: University Kansas Press, 1998), 179–180.

26. Michael Schaller and George Rising, *The Republican Ascendancy: American Politics, 1968–2001* (Wheeling, Ill.: Harlan Davidson, 2002), 59–61; Dolan quoted on p. 63.

27. Interview, Author with Jerry Falwell, July 24, 2003.

28. Diary of Russell Train, 1978 Journal, January 1, 1978, PPRT, LC; Diary of Russell Train, 1977 Journal, March 2, 1977, December 15, 1979, PPRT, LC.

29. Diary of Russell Train, 1978 Journal, January 19, 1978, May 20, 1978, July 2, 1978, PPRT, LC.

30. Diary of Russell Train, 1977 Journal, February 2, 1977, May 24, 1977, June 14, 1977, November 9, 1977, December 12, 1977, PPRT, LC; Letter, Gus Speth to Russell Train, December 8, 1980, PPRT, LC.

31. Diary of Russell Train, 1977 Journal, November 9, 1977, PPRT, LC; Andrus quoted on November 17, 1977; Letter, Russell Train et al. to Jimmy Carter, June 24, 1977, PPRT, LC; Train's views are covered in Council on Environmental Quality, *Response to Committee Questions, Committee on Interior and Insular Affairs, U.S. Senate, 95th Congress, 1st Session, January, 1977* (Washington, D.C.: U.S. Government Printing Office, 1977).

32. Letter, Jimmy Carter to Russell Train, April 23, 1977, PPRT, LC; Press Release, NATO Committee on the Challenges of Modern Society, April 2, 1982, PPRT, LC; Letter, Russell Train to Stuart Eisenstat, May 29, 1979, PPRT, LC.

33. Executive Orders 11990 and 11988; the best survey of Carter's environmental record is Stine, "Environmental Policy during the Carter Presidency."

34. P.L. 96–487 and P.L. 96–510, respectively; Diary of Russell Train, 1977 Journal, August 3, 1977, PPRT, LC.

35. Council on Environmental Quality and the U.S. Department of State, *The Global 2000 Report to the President: Entering the Twenty-First Century* (Washington, D.C.: U.S. Government Printing Office, 1980); Interview, Author with Russell Train, January 9, 2004.

36. Diary of Russell Train, 1978 Journal, December 30, 1978, PPRT, LC; Train apparently forgot Bush's son Jeb.

37. Letter, George Bush to Russell Train, April 25, 1979, PPRT, LC; Letter, William Draper to Russell Train, January 25, 1980, PPRT, LC; Letter, S. S. Livingstone to Russell Train, September 13, 1979, PPRT, LC, Letter, George Bush to Russell Train, May 13, 1980, PPRT, LC.

38. Letter, Nathaniel Reed to George Bush, Russell Train and Fitzgerald Bemiss, April 11, 1979, PPRT, LC.

39. Campaign Press Release, George Bush for President, "An Environmental Policy for the 1980s," PPRT, LC; Letter, Russell Train to Warren Anderson, March 3, 1980, PPRT, LC.

40. John Micklethwait and Adrian Wooldridge, *The Right Nation: Conservative Power in America* (New York, N.Y.: Penguin, 2004), 33; Jon Roper, *The American Presidents: Heroic Leadership from Kennedy to*

Clinton (Chicago, Ill.: Fitzroy Dearborn Publishers, 2000), 159; Joint Statement of the Defenders of Wildlife, the Humane Society, and the Sierra Club before the Republican Party Platform Committee, New York City, 1980, PPRT, LC; Letter, Russell Train to John Studebaker, undated, PPRT, LC; Diary of Russell Train, 1980 Journal, July 14, 1980, PPRT, LC.

41. Letter, Russell Train to George Bush, January 28, 1980, PPRT, LC; Letter, George Bush to Aileen and Russell Train, January 25, 1980, PPRT, LC; Diary of Russell Train, 1980 Journal, October 29, 1980, PPRT, LC.

42. For the 1980 election, see Robert Scheer, *What Happened?: The Story of the 1980 Election* (New York, N.Y.: Random House, 1981); Diary of Russell Train, 1980 Journal, November 4, 1980, November 16, 1980, December 3, 1980, December 4, 1980, PPRT, LC.

43. Interview, Author with Aileen Train, January 9, 2004; Diary of Russell Train, 1981 Journal, January 6, 1981, PPRT, LC.

44. Diary of Russell Train, 1980 Journal, January 20, 1980, PPRT, LC.

45. Ibid., February 25, 1981, PPRT, LC; Dinesh D'Souza, *Ronald Reagan* (New York, N.Y.: Free Press, 1997), 240–241.

46. Garry Wills, *Reagan's America: Innocents at Home* (Garden City, N.Y.: Doubleday, 1987), 346; Diary of Russell Train, 1981 Journal, March 5, 1981, March 10, 1981, PPRT, LC.

47. Diary of Russell Train, 1981 Journal, June 22, 1981, PPRT, LC.

48. Ibid., August 27, 1981, PPRT, LC; Deborah Hart Strober and Gerald S. Strober, *Reagan: The Man and His Presidency* (New York, N.Y.: Houghton Mifflin, 1998), 134.

49. Haynes Johnson, *Sleepwalking through History: America in the Reagan Years* (New York, N.Y.: Norton, 1991), 170; Richard N. L. Andrews, *Managing the Environment, Managing Ourselves: A History of American Environmental Policy* (New Haven, Conn.: Yale University Press, 1999), 258; Gorsuch later married Bureau of Land Management director Robert Burford.

50. John Opie, *Nature's Nation: An Environmental History of the United States* (New York, N.Y.: Harcourt Brace, 1998), 449; Diary of Russell Train, 1981 Journal, April 4, 1981, PPRT, LC; Memorandum for the Files, Russell Train, September 9, 1983, PPRT, LC.

51. Reagan quoted in Strober and Strober, *Reagan,* 70–71, and in Kiron Skinner, Annelise Anderson, and Martin Anderson, *Reagan: A Life in Letters* (New York, N.Y.: Free Press, 2003), 49, 397.

52. Andrews, *Managing the Environment, Managing Ourselves,* 259, 314–315; Stockman quoted on p. 257.

53. Jeffrey K. Stine, "Natural Resources and Environmental Policy," in W. Elliot Brownlee and Hugh Davis Graham, eds., *The Reagan Presidency: Pragmatic Conservatism and Its Legacies* (Lawrence: University of Kansas Press, 2003), 250.

54. Andrews, *Managing the Environment, Managing Ourselves,* 257, 260; Samuel Hays, *Beauty, Health, and Permanence: Environmental Politics in the United States, 1955–1985* (New York, N.Y.: Cambridge University Press, 1987), 504; Letter, Russell Train to James Baker, January 26, 1981, PPRT, LC; Letter, James Baker to Russell Train, February 20, 1981, PPRT, LC.

55. Diary of Russell Train, 1981 Journal, March 10, 1981, PPRT, LC; Train, *Politics, Pollution, and Pandas,* 262.

56. Diary of Russell Train, 1982 Journal, January 6, 1982, PPRT, LC.

57. Ibid., March 2, 1982, PPRT, LC; *Washington Post,* February 2, 1982, 26.

58. *New York Times,* February 7, 1982, 7; subcommittee report quoted in Johnson, *Sleepwalking through History,* 171; Watt quoted in Irwin Unger, *Recent America* (Upper Saddle River, N.J.: Prentice Hall, 2002), 221; Lavelle was later convicted and served six months in prison.

59. Interview, Author with William Ruckelshaus, August 25, 2004; Diary of Russell Train, 1983 Journal, March 17, 1983, PPRT, LC.

60. Memorandum for the Files, Russell Train, November 22, 1983, PPRT, LC; Hays, *Beauty, Health, and Permanence,* 521–522; Opie, *Nature's Nation,* 449.

61. Diary of Russell Train, 1982 Journal, February 2, 1982, PPRT, LC; Diary of Russell Train, 1983 Journal, March 10, 1983, March 21, 1983, PPRT, LC.

62. Interview, Author with William Ruckelshaus, August 25, 2004; Marc Landy, Marc Roberts, and Stephen Thomas, *The Environmental Protection Agency: Asking the Wrong Questions; From Nixon to Clinton* (New York, N.Y.: Oxford University Press, 1994), 251.

63. Hays, *Beauty, Health, and Permanence,* 522–523; Train, *Politics, Pollution, and Pandas,* 133, 269; Memorandum for the Files, Russell Train, January 5, 1983, PPRT, LC.

64. Letter, Russell Train to George Bush, February 11, 1982, PPRT, LC; Diary of Russell Train, 1982 Journal, March 8, 1982, PPRT, LC; Interview, Author with Russell Train, January 9, 2004.

65. Diary of Russell Train, 1982 Journal, March 10, 1982, PPRT, LC.

66. Hal K. Rothman, *The Greening of a Nation?: Environmentalism in the United States since 1945* (Fort Worth, Tex.: Harcourt Brace, 1998), 190; Hays, *Beauty, Health, and Permanence,* 491–492.

67. *New York Times,* October 4, 1984, 26; Interview, Author with Russell Train, January 9, 2004; Train, *Politics, Pollution, and Pandas,* 294–295; *Manila Bulletin,* November 18, 1988, 21.

68. Memorandum for the Files, Russell Train, July 9, 1980, PPRT, LC; Letter, Russell Train to Betty Leslie-Melville, January 26, 1981, PPRT, LC; Reilly quoted in Interview, Author with Russell Train, January 9, 2004.

69. Memorandum for the Files, Russell Train, December 19, 1988, PPRT, LC; Philip quoted in Diary of Russell Train, 1982 Journal, September 20, 1982, PPRT, LC; Interview, Author with Russell Train, January 9, 2004.

70. Memorandum for the Files, Russell Train, July 23, 1979, PPRT, LC; Letter, Russell Peterson to Russell Train, July 17, 1979, PPRT, LC.

71. Diary of Russell Train, 1978 Journal, May 15, 1978, PPRT, LC; Memorandum for the Files, Russell Train, March 13, 1984, PPRT, LC; Memorandum for the Files, Russell Train, January 6, 1983, PPRT, LC; Memorandum for the Files, Russell Train, November 9, 1981, PPRT, LC.

72. Trip Report, Russell Train, February, 1987, PPRT, LC; Trip Report, Russell Train, August, 1987, PPRT, LC.

73. Memorandum for the Files, Russell Train, March 24, 1981, PPRT, LC; Diary of Russell Train, 1981 Journal, January 3–4, 1981, June 24, 1981, PPRT, LC; Interview, Author with Aileen Train, January 9, 2004; Letter, Jane Goodall to Russell Train, June 2, 1986, PPRT, LC; Diary of Russell Train, 1982 Journal, September 20, 1982, PPRT, LC.

74. World Commission on Environment and Development, *Our Common Future* (New York, N.Y.: Oxford University Press, 1987); Interview, Author with Russell Train, January 9, 2004; Train, *Politics, Pollution, and Pandas,* 274–275.

75. *Washington Post,* July 25, 1988, 10; Letter, Russell Train to General Daniel Sherlock, December 15, 1988, PPRT, LC; Draft Conclusions, Commission on Base Closure and Realignment, December 13, 1988, PPRT, LC; Letter, Russell Train to General Dorm Starry, January 26, 1989, PPRT, LC.

76. Letter, Mindly L. Berry to Russell Train, October 7, 1988, PPRT, LC; Letter, Loret Miller Ruppe to Coalition Members, October 22, 1988, PPRT, LC; Bush-Quayle '88 Press Release, Vice President Announces the National Leadership of Conservationists for Bush, October 11, 1988, PPRT, LC.

77. Letter, John Busterud to Russell Train, November 8, 1988, PPRT, LC; Memorandum, Ann Boren to Leadership Committee, Conservationists for Bush, October 12, 1988, PPRT, LC; Membership List, National Leadership Committee, Conservationists for Bush, PPRT, LC; Letter, Douglas Wheeler to Jim Maddox, August 15, 1988, PPRT, LC; Memorandum, Frank Blake to Leadership Committee, Conservationists for Bush, undated, PPRT, LC.

78. *Washington Post,* September 1, 1988, 1, 17; Transcript, Statement of Russell Train before the Republican National Platform Committee, August 8, 1988, PPRT, LC; Letter, Russell Train to George Bush, March 21, 1988, PPRT, LC; Letter, Russell Train to George Bush, September 12, 1988, PPRT, LC; Train, *Politics, Pollution, and Pandas,* 269.

79. *Conservation Foundation Newsletter* (October 1989): 1–7; Letter, Russell Train to George Bush, October 31, 1988, PPRT, LC.

EPILOGUE

1. Ralph Waldo Emerson, *Society and Solitude: Twelve Chapters* (Boston, Mass.: Fields, Osgood and Co., 1870), 173.

2. P.L. 101–549; Interview, Author with Russell Train, January 9, 2004; Richard N. L. Andrews, *Managing the Environment, Managing Ourselves: A History of American Environmental Policy* (New Haven, Conn.: Yale University Press, 1999), 271, 333; Russell E. Train, *Politics, Pollution, and Pandas* (Washington, D.C.: Island Press, 2003), 270.

3. Executive Order 12761; Letter, Russell Train to Bill Clinton, March 8, 1993, PPRT, LC; Letter, Michael Deland to Russell Train, December 19, 1991, PPRT, LC; Council on Environmental Quality, *Council on Environmental Quality Twentieth Anniversary Report, 1990* (Washington, D.C.: U.S. Government Printing Office, 1990).

4. Letter, Russell Train to Robert Bass, February 5, 1988, PPRT, LC; Letter, William Penn Mott to Russell Train, May 8, 1988, PPRT, LC; Letter, Russell Train to Michel Batisse, January 21, 1993, PPRT, LC; Trip Report, Russell Train to Kathryn S. Fuller, June 25, 1990, PPRT, LC; Trip Report, Russell Train to Board of Directors and Members of National Council, September 20, 1991, PPRT, LC; Trip Report, Russell Train to Members of the Board of National Council, April 20, 1994, PPRT, LC; Diary of Russell Train, 1994 Antarctica Trip, PPRT, LC.

5. Letter, Russell Train to His Royal Highness, Duke of Edinburgh, October 22, 1990, PPRT, LC; Interview, Author with Russell Train, January 9, 2004.

6. Interview, Author with Gilbert Omenn, January 21, 2005; Press Release, WWF, February 27, 1991, PPRT, LC; Mission Statement, National Commission on the Environment, 1991, PPRT, LC; Memorandum, Amy Salzman to Russell Train, February 19, 1992, PPRT, LC; Directory, NCE, November 11, 1999, PPRT, LC; Letter, Russell Train to Hazel O'Leary, December 22, 1992, PPRT, LC; Train, *Politics, Pollution, and Pandas,* 317–318.

7. Paarlberg, Robert L., "A Domestic Dispute: Clinton, Congress and International Environmental Policy," *Environment* 38, no. 8 (1996): 16; quoted in John Micklethwait and Adrian Wooldridge, *The Right Nation: Conservative Power in America* (New York, N.Y.: Penguin, 2004), 98.

8. John Opie, *Nature's Nation: An Environmental History of the United States* (New York, N.Y.: Harcourt Brace, 1998), 482–483; Andrews, *Managing the Environment, Managing Ourselves,* 333; Bush quoted in Russell Train, *A Memoir* (Washington, D.C.: self-published, 2000), 348–349.

9. Train, *A Memoir,* 348-349; Sharon Begley, "The Grinch of Rio," *Newsweek* (June 13, 1992): 31; the Biodiversity Convention obligated signatory nations to protect species within their borders and to share profits derived from protected resources.

10. Memo, Mollie Shields to Russell Train, August 21, 1992, PPRT, LC; Robert S. Devine, *Bush versus the Environment* (New York, N.Y.: Anchor Books, 2004), 109, 173; Transcript, Statement of Kathryn S. Fuller, WWF, September 15, 1993, PPRT, LC.

11. Marc Landy, Marc Roberts, and Stephen Thomas, *The Environmental Protection Agency: Asking the Wrong Questions; From Nixon to Clinton* (New York, N.Y.: Oxford University Press, 1994), 295, 301; Daniel J. Fiorino, *Making Environmental Policy* (Berkeley: University of California Press, 1995), 74; David Brower, *The Life and Times of David Brower* (Salt Lake City, Utah: Peregrine Smith Books, 1990), 442.

12. Warren B. Rudman, *Combat: Twelve Years in the U.S. Senate* (New York, N.Y.: Random House, 1996), 243; Train, *Politics, Pollution, and Pandas,* 316.

13. Interview, Author with Tom Clausen, January 20, 2005.

14. Jim Wright, *Balance of Power* (Atlanta, Ga.: Turner Publishing, 1996), 490.

15. DeLay quoted in Michael Schaller and George Rising, *The Republican Ascendancy: American Politics, 1968–2001* (Wheeling, Ill.: Harlan Davidson, 2002), 132.

16. National Commission on the Environment, *Choosing a Sustainable Future: The Report of the National Commission on the Environment* (Washington, D.C.: Island Press, 1993); Interview, Author with Russell Train, January 9, 2004; Albert Gore, *Earth in the Balance: Ecology and the Human Spirit* (New York, N.Y.: Penguin, 1993); *Los Angeles Times,* December 12, 1992, 4.

17. Michael Weisskopf, "The Dream Team?" *Outside* (November 1992): 74; Memo, EPA Office of Policy Planning and Evaluation to Staff, January 28, 1992, PPRT, LC.

18. P.L. 103–433; *Business Week* (December 12, 1994): 17; Letter, Russell Train to Max Baucus, March 30, 1993, PPRT, LC; Interview, Author with Michael McCloskey, June 29, 2004.

19. Interview, Author with Joe Browder, June 29, 2004.

20. Invitation, "Great American: A Salute to Russell E. Train," September 20, 1994, PPRT, LC; List of Attendees, Russell Train Retirement Dinner, September 20, 1994, PPRT, LC; Letter, Max Baucus to Russell Train, October 17, 1994, PPRT, LC; Letter, George Bush to Russell Train, September 20, 1994, PPRT, LC; Letter, Peter Berle to Russell Train, September 20, 1994, PPRT, LC; Letter, Perez Olindo to Russell Train, July 18, 1994, PPRT, LC; Interview, Author with Russell Train, January 9, 2004.

21. Reilly quoted in Brochure, The Russell E. Train Education for Nature Fund, PPRT, LC; Interview, Author with Russell Train, January 17, 2004.

22. Letter, Russell Train to George O'Neill, November 16, 1994, PPRT, LC; Train, *A Memoir,* 358–364.

23. *Sierra Club Bulletin* 89, no. 5 (September–October 2004): 37–51; Interview, Author with Michael McCloskey, June 29, 2004.

24. Interview, Author with Alice Rivlin, January 24, 2005; see Devine, *Bush versus the Environment;* William Snape and John M. Carter, *Weakening the National Environmental Policy Act: How the Bush Administration Uses the Judicial System to Weaken Environmental Protections* (Washington, D.C.: Defenders of Wildlife, 2003); Environmental Integrity Project, *Paying Less to Pollute* (Washington, D.C.: Environmental Integrity Project, 2003); Rick Abraham, *The Dirty Truth: George Bush's Oil and Chemical Dependency* (Houston, Tex.: Mainstream Press, 2000); *Washington Post,* June 6, 2003, 7, July 1, 2003, 6.

25. *Dallas Morning News,* December 23, 2004; Transcript, Interview, Diane Rehm with Russell Train, December 11, 2003, National Public Radio; Christine Todd Whitman, *It's My Party Too: The Battle for the Heart of the GOP and the Future of America* (New York, N.Y.: Penguin, 2005).

26. Interview, Diane Rehm with Russell Train, December 11, 2003, National Public Radio.

27. *Baltimore Sun,* October 20, 2004, 18; *Minneapolis Star Tribune,* September 15, 2004, 10; *St. Paul Pioneer Press,* September 15, 2004; *Seattle Post-Intelligencer,* September 8, 2004, 2; Memorandum for the Files, Russell Train, November 11, 2004, PPRT, LC.

28. Train, *Politics, Pollution, and Pandas,* 324.

SELECT BIBLIOGRAPHY

PRIMARY SOURCES

Archival Material

Edmund Muskie Archives, Bates College, Lewiston, Maine
    United States Senate Staff Office Files

Personal Papers of Russell E. Train (The personal papers of Russell Train are in the Manuscript Division of the Library of Congress, Washington, D.C., but remain uncataloged. They consist of letters, office memoranda, citations, photographs, trip journals, and his daily diary.)

Richard M. Nixon Presidential Materials Project, National Archives II, College Park, Maryland
    White House Special Files
        Charles Colson Files
        Egil Krogh Files
        H. R. Haldeman Files
        Presidential Task Forces Files
        President's Handwriting Files
        President's Office Files
    White House Central Files
        Edward David Files
        Edwin Harper Files
        Environmental Protection Agency Files
        Health Files
        John Whitaker Files
        Natural Resources Files
        NIPCC (National Industrial Pollution Control Council) Files
        PACEO (President's Advisory Council on Executive Organizations) Files
        Proposed Departments Files
        Transition Task Forces Files

Interviews

Philip Berry
Leon Billings
Joe Browder
David Brower
Tom Clausen
Jerry Falwell
Richard Fairbanks
George Hartzog
Christian Herter
Walter Hickel
Michael McCloskey
Laurence Moss
Don Nicholl
Gilbert Omenn
William Reilly
Alice Rivlin
William Ruckelshaus
Maurice Stans
Aileen Train
Russell Train
John Whitaker

Government Documents

Bureau of the Census. *Historical Statistics of the United States: Colonial Times to 1970.* 2 vols. Washington, D.C.: U.S. Government Printing Office, 1975.
Congressional Record. 91st Cong., 1st sess., January 1969, vol. 115, pt. 2.
———. 92nd Cong., 2nd sess., June 1972, vol. 118, pt. 6.
———. 93rd Cong., 1st sess., September 1973, vol. 119, pts. 26–27.
———. 93rd Cong., 2nd sess., June 1974, vol. 120, pt. 10.
Council on Environmental Quality. *Coal Surface Mining and Reclamation: An Environmental and Economic Assessment of Alternatives.* Washington, D.C.: U.S. Government Printing Office, 1973.
———. *Council on Environmental Quality Twentieth Anniversary Report, 1990.* Washington, D.C.: U.S. Government Printing Office, 1990.
———. *First Annual Report of the President's Council on Environmental Quality.* Washington, D.C.: U.S. Government Printing Office, 1970.
———. *The 1973 Annual Report of the President's Council on Environmental Quality.* Washington, D.C.: U.S. Government Printing Office, 1973.

————. *Ocean Dumping: A National Policy.* Washington, D.C.: U.S. Government Printing Office, 1970.

————. *The President's 1971 Environmental Program.* Washington, D.C.: U.S. Government Printing Office, 1971.

————. *Response to Committee Questions, Committee on Interior and Insular Affairs, U.S. Senate, 95th Congress, 1st Session, January, 1977.* Washington, D.C.: U.S. Government Printing Office, 1977.

————. *Second Annual Report of the President's Council on Environmental Quality.* Washington, D.C.: U.S. Government Printing Office, 1971.

Council on Environmental Quality and the U.S. Department of State. *The Global 2000 Report to the President: Entering the Twenty-First Century.* Washington, D.C.: U.S. Government Printing Office, 1980.

Public Land Law Review Commission. *One Third of the Nation's Land: A Report to the President and Congress.* Washington, D.C.: U.S. Government Printing Office, 1970.

Public Papers of the Presidents, Richard Nixon. Vols. 1–5, 1969–1974. Washington, D.C.: U.S. Government Printing Office, 1969.

United States Congress. Senate Committee on Commerce. *Report of the Senate Committee on Commerce on S.3149, Toxic Substances Control Act, March 16, 1976.* Washington, D.C.: U.S. Government Printing Office, 1976.

United States Congress. Senate Committee on Interior and Insular Affairs. *Hearings before the Senate Committee on Interior and Insular Affairs on S.1075, S.237 and S.1725.* 91st Cong., 1st sess., April 16, 1969.

————. *Hearings before the Senate Committee on Interior and Insular Affairs on the Nomination of Russell E. Train to the Council on Environmental Quality,* 91st Cong., 2nd sess., February 5, 1970.

————. *Hearings on S.1075 before the Senate Committee on Interior and Insular Affairs.* 91st Cong., 1st sess., June 12–16, 1969.

————. *Hearings on S.1401, A Bill to Amend the Land and Water Conservation Fund Act.* 90th Cong., 2nd sess., February 8, 1968.

————. *Hearings on the Nomination of Russell E. Train to be Under Secretary of Interior.* 91st Cong., 1st sess., February 4, 1969.

United States Congress. Senate Committee on Public Works. *Hearings on the Compliance with Title II (Auto Emission Standards) of the Clean Air Act.* 93rd Cong., 1st sess., November 6, 1973.

————. *Hearings on the Nomination of Russell E. Train to be Administrator of the Environmental Protection Agency.* 93rd Cong., 1st sess., August 1, 1973.

United States Congress. Senate Subcommittee on Merchant Marine and Fisheries. *Hearings on S.2984, A Bill to Prevent Importation of Endangered Species of Fish and Wildlife into the United States.* 90th Cong., 2nd sess., July 24, 1968.

United States Department of Agriculture. *Report on the PCB Incidence in the Western United States.* Washington, D.C.: U.S. Government Printing Office, 1980.

United States National Water Commission. *Annual Report, 1969.* Washington, D.C.: U.S. Government Printing Office, 1969.

Memoirs

Brower, David. *The Life and Times of David Brower.* Salt Lake City, Utah: Peregrine Smith Books, 1990.

Carter, Jimmy. *Keeping Faith: Memoirs of a President.* New York, N.Y.: Bantam, 1982.

Ehrlichman, John. *Witness to Power: The Nixon Years.* New York, N.Y.: Pocket Books, 1982.

Eisenhower, Julie Nixon. *Pat Nixon: The Untold Story.* New York, N.Y.: Simon and Schuster, 1986.

Ford, Betty. *The Times of My Life.* New York, N.Y.: Harper and Row, 1978.

Ford, Gerald. *A Time to Heal: The Autobiography of Gerald R. Ford.* New York, N.Y.: Harper and Row, 1979.

Graham, Katharine. *Katharine Graham's Washington.* New York, N.Y.: Random House, 2002.

———. *Personal History.* New York, N.Y.: Alfred A. Knopf, 1997.

Greenfield, Meg. *Washington.* New York, N.Y.: Public Affairs, 2001.

Hickel, Walter. *Who Owns America?* Englewood Cliffs, N.J.: Prentice Hall, 1971.

Kissinger, Henry. *White House Years.* Boston, Mass.: Little, Brown, 1979.

———. *Years of Upheaval.* Boston, Mass.: Little, Brown, 1982.

Nixon, Richard. *RN: The Memoirs of Richard Nixon.* New York, N.Y.: Grosset and Dunlap, 1978.

Peterson, Merrill, ed. *Thomas Jefferson: Writings.* New York, N.Y.: Literary Classics of the United States, 1984.

Quarles, John. *Cleaning Up America: An Insider's View of the Environmental Protection Agency.* Boston, Mass.: Houghton Mifflin, 1976.

Rudman, Warren B. *Combat: Twelve Years in the U.S. Senate.* New York, N.Y.: Random House, 1996.

Train, Russell. *The Bowdoin Family.* Washington, D.C.: self-published, 2000.

———. *A Memoir.* Washington, D.C.: self-published, 2000.

———. *Politics, Pollution, and Pandas.* Washington, D.C.: Island Press, 2003.

———. *The Train Family.* Washington, D.C.: self-published, 2000.

Whitaker, John. *Striking a Balance: Environment and Natural Resources Policy in the Nixon-Ford Years.* Washington, D.C.: American Enterprise Institute, 1976.

Whitman, Christine Todd. *It's My Party Too: The Battle for the Heart of the GOP and the Future of America.* New York, N.Y.: Penguin, 2005.

Wright, Jim. *Balance of Power.* Atlanta, Ga.: Turner Publishing, 1996.

Environmental Publications

National Commission on the Environment, *Choosing a Sustainable Future: The Report of the National Commission on the Environment.* Washington, D.C.: Island Press, 1993.

Scott, Peter, ed. *The Launching of the New Ark: The First Report of the President and Trustees of the World Wildlife Fund*. London, England: Collins, 1965.
World Wildlife Fund–United States. *Annual Report, 1978*. Washington, D.C.: World Wildlife Fund, 1978.
World Wildlife Fund–United States. *Annual Report, 2003*. Washington, D.C.: World Wildlife Fund, 2003.

SECONDARY SOURCES

Abraham, Rick. *The Dirty Truth: George Bush's Oil and Chemical Dependency*. Houston, Tex.: Mainstream Press, 2000.
Aitken, Jonathan. *Nixon: A Life*. Washington, D.C.: Regnery Publishing, 1993.
Alexander, Charles C. *Holding the Line: The Eisenhower Era, 1952–1961*. Bloomington: Indiana University Press, 1975.
Ambrose, Stephen. *Nixon: The Triumph of a Politician, 1962–1972*. New York, N.Y.: Simon and Schuster, 1989.
———. *Nixon: Ruin and Recovery, 1972–1990*. New York, N.Y.: Simon and Schuster, 1991.
Anderson, Frederick R., and Robert H. Daniels. *NEPA in the Courts*. Baltimore, Md.: Johns Hopkins University Press, 1973.
Andrews, Richard N. L. *Environmental Policy and Administrative Change*. Lexington, Mass.: D.C. Heath, 1976.
———. *Managing the Environment, Managing Ourselves: A History of American Environmental Policy*. New Haven, Conn.: Yale University Press, 1999.
Applewhite, Ashton, et al. *And I Quote*. New York, N.Y.: St. Martin's Press, 2003.
Bench, Edward L. *The United States Navy: 200 Years*. New York, N.Y.: Holt, 1986.
Blake, Robert. *Disraeli*. London, England: Eyre and Spottiswoode, 1966.
Bosselman, Fred, and David Callies. *The Quiet Revolution in Land Use Control*. Washington, D.C.: U.S. Government Printing Office, 1971.
Brands, H. W. *The Strange Death of American Liberalism*. New Haven, Conn.: Yale University Press, 2001.
Brown, Ruth Murray. *For a Christian Nation: A History of the Religious Right*. Amherst, N.Y.: Prometheus Books, 2002.
Bundy, William. *A Tangled Web: The Making of Foreign Policy in the Nixon Presidency*. New York, N.Y.: Hill and Wang, 1998.
Carson, Rachel. *Silent Spring*. Boston, Mass.: Houghton Mifflin, 1962.
Caute, David. *The Year of the Barricades*. New York, N.Y.: Harper and Row, 1988.
Chafe, William C. *The Unfinished Journey*. New York, N.Y.: Oxford University Press, 1986.
Clary, David. *Timber and the Forest Service*. Lawrence: University of Kansas Press, 1986.
Coates, Peter. *The Trans-Alaska Pipeline Controversy*. Bethlehem, Pa.: Lehigh University Press, 1991.

Congressional Quarterly. *Gerald Ford: The Man and His Record*. Washington, D.C.: Congressional Quarterly, Inc., 1974.

Congressional Quarterly Almanac. 93rd Cong., 2nd sess., 1974, vol. 30.

Congressional Research Service. *A Legislative History of the Water Pollution Control Act Amendments of 1972*. Washington, D.C.: U.S. Government Printing Office, 1973.

Crowley, Monica. *Nixon in Winter*. New York, N.Y.: Random House, 1998.

Culhane, Paul J. *Public Lands Politics*. Baltimore, Md.: Johns Hopkins University Press, 1981.

Devine, Robert S. *Bush versus the Environment*. New York, N.Y.: Anchor Books, 2004.

D'Souza, Dinesh. *Ronald Reagan*. New York, N.Y.: Free Press, 1997.

Dunlap, Thomas. *Saving America's Wildlife*. Princeton, N.J.: Princeton University Press, 1988.

Easton, Robert. *Black Tide: The Santa Barbara Oil Spill and Its Consequences*. New York, N.Y.: Delacorte Press, 1972.

Edwards, Lee. *The Conservative Revolution: The Movement that Remade America*. New York, N.Y.: Free Press, 1999.

Emerson, Ralph Waldo. *Society and Solitude: Twelve Chapters*. Boston, Mass.: Fields, Osgood and Co., 1870.

Emery, Fred. *The Corruption of American Politics and the Fall of Richard Nixon*. New York, N.Y.: Touchstone, 1994.

Environmental Integrity Project. *Paying Less to Pollute*. Washington, D.C.: Environmental Integrity Project, 2003.

Fiorino, Daniel J. *Making Environmental Policy*. Berkeley: University of California Press, 1995.

Fischer, David Hackett. *Albion's Seed: Four British Folkways in America*. New York, N.Y.: Oxford University Press, 1989.

Flippen, J. Brooks. *Nixon and the Environment*. Albuquerque: University of New Mexico Press, 2000.

Galbraith, John Kenneth. *The Affluent Society*. Boston, Mass: Houghton Mifflin, 1958.

Glad, Betty. *In Search of the Great White House*. New York, N.Y.: Norton, 1980.

Gore, Albert. *Earth in the Balance: Ecology and the Human Spirit*. New York, N.Y.: Penguin, 1993.

Gould, Lewis L. *Lady Bird Johnson and the Environment*. Lawrence: University of Kansas Press, 1988.

Greenstein, Fred. *The Hidden Hand Presidency: Eisenhower as Leader*. Baltimore, Md.: Johns Hopkins University Press, 1994.

Hallet, Robin. *Africa since 1875*. Ann Arbor: University of Michigan Press, 1974.

Hand, Susan Train. *John Trayne and Some of His Descendants*. New York, N.Y.: self-published, 1933.

Harvey, Mark. *A Symbol of Wilderness: Echo Park and the American Conservation Movement*. Albuquerque: University of New Mexico Press, 1994.

Hays, Samuel P. *Beauty, Health, and Permanence: Environmental Politics in the United States, 1955–1985*. New York, N.Y.: Cambridge University Press, 1987.

———. *Conservation and the Gospel of Efficiency: The Progressive Conservation Movement, 1890–1920.* Cambridge, Mass: Harvard University Press, 1959.

Hite, James C., and James M. Stepp. *Coastal Zone Resource Management.* New York, N.Y.: Praeger, 1971.

Hoole, Francis, Robert Friedheim, and Timothy Hennessey, eds. *Making Ocean Policy.* Boulder, Colo.: Westview Press, 1981.

Horwitch, Melvin. *Clipped Wings: The American SST Conflict.* Cambridge, Mass.: MIT Press, 1982.

Johnson, Donald, and Kirk Porter. *National Party Platforms.* Urbana: University of Illinois Press, 1973.

Johnson, Haynes. *Sleepwalking through History: America in the Reagan Years.* New York, N.Y.: Norton, 1991.

Kearns, Doris. *Lyndon Johnson and the American Dream.* New York, N.Y.: Harper and Row, 1976.

Kehoe, Terence. *Cleaning Up the Great Lakes: From Cooperation to Confrontation.* DeKalb: Northern Illinois University Press, 1997.

Kennedy, David M. *Freedom from Fear: The American People in Depression and War, 1929–1945.* New York, N.Y.: Oxford University Press, 1999.

Kennon, Donald R., and Rebecca M. Rogers. *The Committee on Ways and Means: A Bicentennial History, 1789–1989.* Washington, D.C.: U.S. Government Printing Office, 1989.

Kristol, Irving, ed. *Neo-Conservatism: The Autobiography of an Idea; Selected Essays.* New York, N.Y.: Free Press, 1995.

Kutler, Stanley. *Abuse of Power.* New York, N.Y.: Free Press, 1997.

———. *The Wars of Watergate.* New York, N.Y.: Alfred A. Knopf, 1990.

Landy, Marc, Marc Roberts, and Stephen Thomas. *The Environmental Protection Agency: Asking the Wrong Questions; From Nixon to Clinton.* New York, N.Y.: Oxford University Press, 1994.

Lear, Linda. *Rachel Carson: Witness for Nature.* New York, N.Y.: Holt, 1997.

Lieber, Harvey. *Federalism and Clean Waters: The 1972 Water Pollution Control Act.* Lexington, Mass.: D.C. Heath and Co., 1975.

Link, Arthur S., and William B Catton. *American Epoch.* New York, N.Y.: Alfred A. Knopf, 1980.

Lippman, Theo, Jr., and Donald C. Hansen. *Muskie.* New York, N.Y.: Norton, 1971.

Mahan, Alfred Thayer. *The Influence of Seapower on History.* Boston, Mass.: Little, Brown, 1890.

Mason, Robert. *Richard Nixon and the Quest for a New Majority.* Chapel Hill: University of North Carolina Press, 2004.

Mayer, Carl, and George A. Riley. *Public Domain, Private Dominion.* San Francisco, Calif.: Sierra Club Books, 1985.

Mazlish, Bruce, and Edwin Diamond. *Jimmy Carter: A Character Portrait.* New York, N.Y.: Simon and Schuster, 1979.

McEvoy, James. "The American Concern for the Environment." In *Social Behavior, Natural Resources, and the Environment,* edited by William Burch et al., 214–236. New York, N.Y.: Harper and Row, 1972.

McHarg, Ian L. *Design with Nature.* Garden City, N.Y.: Natural History Press, 1969.

McPhee, John. *Encounters with the Archdruid.* New York, N.Y.: Farrar, Straus and Giroux, 1971.

Micklethwait, John, and Adrian Wooldridge. *The Right Nation: Conservative Power in America.* New York, N.Y.: Penguin, 2004.

Morris, Roger. *Richard Milhous Nixon: The Rise of an American Politician.* New York, N.Y.: Holt, 1990.

Nathan, Richard. *The Plot that Failed: Nixon and the Administrative Presidency.* New York, N.Y.: John Wiley and Sons, 1975.

Nelson Rockefeller: Passionate Millionaire. Videocassette. New York: ABC News Productions, 1998.

Netboy, Anthony. *The Atlantic Salmon: A Vanishing Species?* London, England: Faber, 1968.

Nevin, David. *Muskie of Maine.* New York, N.Y.: Random House, 1972.

Oliver, Roland. *The African Experience.* New York, N.Y.: HarperCollins, 1991.

Olson, Keith. *The Presidential Scandal that Shocked America.* Lawrence: University of Kansas Press, 2003.

Opie, John. *Nature's Nation: An Environmental History of the United States.* New York, N.Y.: Harcourt Brace, 1998.

Osborn, Fairfield. *Our Plundered Planet.* Boston, Mass.: Little, Brown, 1948.

Palmer, Tim. *Endangered Rivers and the Conservation Movement.* Berkeley: University of California Press, 1986.

Parmet, Herbert. *Richard Nixon and His America.* Boston, Mass.: Little, Brown, 1990.

Petulla, Joseph M. *American Environmental History.* San Francisco, Calif.: Boyd and Fraser, 1977.

Price, Daniel. *The 99th Hour: The Population Crisis and the United States.* Chapel Hill: University of North Carolina Press, 1967.

Rathlesberger, James, ed. *Nixon and the Environment.* New York, N.Y.: Village Voice, 1972.

Reichard, Gary. *The Reaffirmation of Republicanism: Eisenhower and the Eighty-Third Congress.* Knoxville: University of Tennessee Press, 1975.

Remini, Robert. *Andrew Jackson and the Course of American Freedom, 1822–1832.* 2 vols. New York, N.Y.: Harper and Row, 1981.

Richardson, Elmo. *Dams, Parks, and Politics: Resource Development and Preservation in the Truman-Eisenhower Era.* Lexington: University of Kentucky Press, 1973.

Roper, Jon. *The American Presidents: Heroic Leadership from Kennedy to Clinton.* Chicago, Ill: Fitzroy Dearborn Publishers, 2000.

Rosenbaum, Walter. *The Politics of Environmental Concern.* New York, N.Y.: Praeger, 1974.

Rothman, Hal K. *The Greening of a Nation?: Environmentalism in the United States since 1945.* Fort Worth, Tex.: Harcourt Brace, 1998.

Russell, Jacqueline. *Thirty-Five Years of Conserving Wildlife in Africa: A History of the African Wildlife Foundation.* Washington, D.C.: African Wildlife Foundation, 1996.

Sansom, Robert L. *The New American Dream Machine: Towards a Simpler Lifestyle in an Environmental Age.* Garden City, N.J.: Anchor Press, 1976.

Scammon, Richard M., and Ben J. Wattenberg. *The Real Majority.* New York, N.Y.: Coward-McCain, 1970.

Schaller, Michael, and George Rising. *The Republican Ascendancy: American Politics, 1968–2001.* Wheeling, Ill.: Harlan Davidson, 2002.

Scheer, Robert. *What Happened?: The Story of the 1980 Election.* New York, N.Y.: Random House, 1981.

Scheffer, Victor B. *The Shaping of Environmentalism in America.* Seattle: University of Washington Press, 1998.

Shanley, Robert A. *Presidential Influence and Environmental Policy.* Westport, Conn.: Greenwood Press, 1991.

Shoup, Lawrence. *The Carter Presidency and Beyond.* Palo Alto, Calif.: Ramparts Press, 1980.

Skinner, Kiron, Annelise Anderson, and Martin Anderson. *Reagan: A Life in Letters.* New York, N.Y.: Free Press, 2003.

Snape, William, and John M. Carter. *Weakening the National Environmental Policy Act: How the Bush Administration Uses the Judicial System to Weaken Environmental Protections.* Washington, D.C.: Defenders of Wildlife, 2003.

Solberg, Carl. *Hubert Humphrey.* New York, N.Y.: Norton, 1984.

Stacks, John F. *Stripping: The Surface Mining of America.* San Francisco, Calif.: Sierra Club Books, 1972.

Stine, Jeffrey K. "Environmental Policy during the Carter Presidency." In *The Carter Presidency: Policy Choices in the Post–New Deal Era,* edited by Gary M. Fink and Hugh Davis Graham. Lawrence: University of Kansas Press, 1998.

———. "Natural Resources and Environmental Policy." In *The Reagan Presidency: Pragmatic Conservatism and Its Legacies,* edited by W. Elliot Brownlee and Hugh Davis Graham. Lawrence: University of Kansas Press, 2003.

Strober, Deborah Hart, and Gerald S. Strober. *Reagan: The Man and His Presidency.* New York, N.Y.: Houghton Mifflin, 1998.

Thornton, Richard C. *The Nixon-Kissinger Years.* St. Paul, Minn.: Paragon, 2001.

Unger, Irwin. *Recent America.* Upper Saddle River, N.J.: Prentice Hall, 2002.

United States House of Representatives, Committee on House Administration. *The Presidential Campaign, 1976.* Washington, D.C.: U.S. Government Printing Office, 1978.

Vileisis, Ann. *Discovering the Unknown Landscape: A History of America's Wetlands.* Washington, D.C.: Island Press, 1997.

Walth, Brent. *Fire at Eden's Gate: Tom McCall and the Oregon Story.* Portland: Oregon Historical Society Press, 1994.

Wills, Garry. *Reagan's America: Innocents at Home.* Garden City, N.Y.: Doubleday, 1987.

Wilson, Carroll W., et al. *Coal—A Bridge to the Future: A Report of the World Coal Study, WO-COL.* Cambridge, Mass.: Ballinger, 1980.

Winks, Robin W. *Laurance S. Rockefeller: Catalyst for Conservation.* Washington, D.C.: Island Press, 1997.

Woodward, Bob. *Five Presidents and the Legacy of Watergate.* New York, N.Y.: Simon and Schuster, 1999.

World Commission on Environment and Development. *Our Common Future.* New York, N.Y.: Oxford University Press, 1987.

INDEX